Chasing Hillary

Chasing Hillary

Ten Years,
Two Presidential Campaigns, and
One Intact Glass Ceiling

Amy Chozick

HARPER

An Imprint of HarperCollins*Publishers*

HarperCollins books may be purchased for educational, business, or sales promotional use. For information, please email the Special Markets Department at SPsales@harpercollins.com.

FIRST EDITION

Photograph on page 377 courtesy of the author.
Designed by Bonni Leon-Berman

Library of Congress Cataloging-in-Publication Data has been applied for.

ISBN 978-0-06-241359-8

18 19 20 21 22 LSC 10 9 8 7 6 5 4 3 2 1

For Bobby

I know of no American who starts from a higher level of aspiration than the journalist. He is, in his first phase, genuinely romantic. He plans to be both an artist and a moralist—a master of lovely words and a merchant of sound ideas. He ends, commonly, as the most depressing jackass in his community—that is, if his career goes on to what is called success. He becomes the repository of all its worst delusions and superstitions. He becomes the darling of all its frauds and idiots, and the despair of all its honest men.

—H. L. MENCKEN

I suppose I could have stayed home and baked cookies and had teas, but what I decided to do was to fulfill my profession, which I entered before my husband was in public life.

—HILLARY CLINTON, 1992

Contents

Author's Note

THIS BOOK IS A WORK OF NONFICTION in that everything in it happened. But this is not a work of journalism, in that the recollections, conversations, and characters are based on my own impressions and memories of covering Hillary Clinton and her family beginning in 2007 and ending with the inauguration of Donald J. Trump on January 20, 2017. I hired a professional fact-checker to review—and scrutinize—my version of events. My story is based on hundreds of interviews that took place during this ten-year period, documented in transcripts, audio recordings, and stacks of reporter's notebooks that I stuffed into plastic containers and kept under my bed just in case I ever wrote a book. I also referred to campaign materials, archival documents, and the Miller Center's oral history of the White House years. I've always kept journals, and even at my most exhausted would scribble down conversations from the campaign trail and my musings about whatever town we were in or news events that unfolded that day. I took lots of photos to help re-create scenes. I changed some names and identifying details, and gave lots of people pseudonyms, sometimes to protect the innocent but usually to protect the story—I think having to remember the names of dozens of political operatives who all essentially perform the same purpose is boring. In the rare cases in which I couldn't confirm exact details or dialogue, I re-created them from memory and, when possible, reviewed them with the people involved. Any material that was initially mutually agreed upon to be off the record was passed on to me by a separate source or used with permission. This book—indeed, my role in it—would not exist without the *Wall Street Journal* and the *New York Times* entrusting me with the Hillary beat, believing in my journalism and springing for me to travel the country to trail the would-be First Woman President.

Chasing Hillary

1

Happy Hillary

Everything was beautiful, and nothing hurt.
—KURT VONNEGUT, *SLAUGHTERHOUSE-FIVE*

NOVEMBER 8, 2016

No one spoke on the press van. I rested my knees on the seat in front of me and sank into the back row looking out the window at the Hudson River. In the past twenty-four hours, I'd slept maybe forty-five minutes and that was by accident. I'd fallen asleep sprawled out longways in an armchair in the lobby of the Ritz-Carlton in White Plains, New York, waiting for her campaign staff to wrangle us back into the press van to go watch Hillary Clinton vote. Ever since Labor Day, we'd basically lived in the slim silver tower that, until Hillary's press corps' arrival, seemed built for the sole purpose of accommodating hedge-fund managers and hookers.

Hillary and Donald Trump both liked to fly back to New York at night so they could sleep in their own beds. The Ritz put the traveling press in proximity to the Clintons' home in Chappaqua while still acquiring Marriott points, which were really the only thing that sustained us in those final months on the road. Entire conversations revolved around Marriott points, how many we had, how we'd cash them in when the campaign came to an end.

I couldn't tell if I was just tired or still had the busy, swirling head of someone who had downed three Dixie cups full of lukewarm champagne before filing my final campaign-trail story for the *New York Times* at around 3:45 a.m. It was probably both.

At first, I'd resisted the leftover champagne that hours earlier made its way from Hillary's front cabin on the "Stronger Together" plane to our rowdy press quarters in the back.

I'd learned my lesson eight years earlier, before I joined the *Times* and adopted my role as detached political reporter. Hillary had walked to the back of her 2008 campaign plane, the Hill Force One, and stretched out a tray of peach cobbler she'd picked up from the Kitchen Express in Little Rock. I heaped a pile of it onto my plate. The image landed in the Associated Press. There I was, a *Wall Street Journal* cub reporter, literally allowing the candidate to feed the press.

But now it was after 2:00 a.m. on Election Day, and it was setting in that it was all over. The traveling press (or Travelers, as the campaign called us) was a pile of emotions and adrenaline. This wasn't just Hillary's victory party. It was ours. We'd made it through 577 days of the most noxious, soul-crushing presidential campaign in modern history. Now we'd get our reward—the chance to cover history, the election of the first woman president, or the FWP as we called her.

The campaign sent the Travelers our final schedule. "After over 120 schedules, 300 meals, and countless Marriott points, we hope you enjoy the day on the road . . ."

White Plains → Pittsburgh → Grand Rapids → Philadelphia →
    Raleigh → White Plains

Until that last day, I hadn't felt as though I was covering a winning campaign. Not that I thought Trump would win. I believed in the data, yet I couldn't shake the nagging on-the-ground sensation that Hillary wouldn't win. In mid-October, after the *Access Hollywood* video landed, I'd been working mostly from the New York office trying to keep up with the dizzying news cycle. I'd asked my editors at the *Times* to send me back out on the road.

"I just feel like the election isn't happening in my cubicle," I pleaded to Very Senior Editor, who—hand raised as if answering a question in science class—reminded me that the *Times*' Upshot election model gave Hillary a 93 percent chance of winning. "But it's over," Very Senior Editor replied.

It was over, and we had to prepare. I put the finishing touches on a thirty-five-hundred-word tome about Hillary's path to the presidency that the *Times* art department had already laid out across six front-page columns under the headline MADAM PRESIDENT. The nut graph, which my coauthor, Patrick Healy, and I had spent weeks perfecting, read:

> No one in modern politics, male or female, has had to withstand more indignities, setbacks and cynicism. She developed protective armor that made the real Hillary Clinton an enigma. But if she was guarded about her feelings and opinions, she believed it was in careful pursuit of a dream for generations of Americans: the election of the country's first woman president.

I had two more stories to finish—one on how Hillary planned to work with Republicans and one on the Hillary Doctrine, foreign and domestic policy. I also had a couple of features in the can, scheduled to run in the *Times*' commemorative women's section the day after the election. Advertisers had already bought space in the historic special edition. I even had a story ready for the paper's Sunday Styles section about how Hillary would be the booziest president since FDR.

> Beset by stereotypes that she is a hall-monitor type, buttoned up and bookish, churchgoing and dutiful, but not much fun at a keg party, in reality, Mrs. Clinton enjoys a cocktail—or three—more than most previous presidents.

I could see everything from where I was sitting. Hillary in the front cabin. Bill, Chelsea, all their aides, standing in the aisles and on their seats. Towers of pizza boxes balanced on turned-down tray tables. The champagne, followed by coffee, that went around to all Hillary's closest aides, the ones from the White House and the State Department, the ones whom she'd pretended to sideline during the campaign—Hillary's soon-to-be West Wing caffeinated and floating at thirty-nine thousand feet. Jon Bon Jovi, a family friend, perched on Hillary's armrest with his guitar, his black jeans practically touching her shoulder.

Even the Secret Service agents, who usually sat stiff-backed in the middle cabin, dividing the press from the candidate, now roamed the plane. A hunky sharpshooter with camouflage pants, a bulletproof vest, and pointy black eyebrows ventured to our cabin to peruse Hillary's almost entirely female press corps.

Over the cacophony of the press cabin—a mix of "Single Ladies" and "Don't Stop Believin'" blasted from a photographer's karaoke machine and a network producer's competing portable speaker—I could hear Hillary's belly laugh. She wore an ample open-mouthed smile.

In ten years of covering Hillary, the formative years of my adult life, really, I'd never seen her so happy. This particular smile, wide and toothy, an O shape that spread over the circumference of her face, I'd seen maybe three other times: on the chilly night in 2008 when she won the New Hampshire primary; in October midway through a late-night flight to Pittsburgh when Tim Kaine, a couple buttons undone and looking like every Catholic housewife's fantasy, sidled up next to her; and that past Saturday when she raised both arms overhead and allowed herself to get soaked under a tropical storm in Pembroke Pines, Florida, throwing caution and her John Barrett blowout to the wind.

But those smiles always faded. This one lasted for twenty-one hours of campaigning and well into Election Day when Hillary stepped out of her black "Scooby van" at Douglas Grafflin Elementary School in Chappy and followed the VOTE HERE/VOTE AQUÍ instructions.

It was a sign of our exhaustion that no one spoke. Usually, the Travelers couldn't shut up. The day before, on the tarmac in White Plains, a heated debate erupted about whether Hillary would wear a gown or one of those embellished satin tunics over wide-legged pants to the inaugural balls.

"Of *course* she's going to wear a dress," somebody argued.

"I don't know. Pants could be revolutionary."

"Yeah, and has she even shown her shoulders since 2009?"

We snapped selfies and talked about our postelection plans—vacations to Italy, the Turks and Caicos, a spa in Arizona (that accepts Marriott points), a juice cleanse. After that, we'd reunite in Washington to cover the FWP in the White House.

Hillary's cadre of protective male press aides—a rotating cast of about half a dozen whom I will collectively refer to as "The Guys" and whose job descriptions included protecting Hillary in the press and dealing with the endless inquiries, requests, and groveling from the reporters who covered her—compared the mood inside the campaign to the final lap of the Tour de France when the wind whips at your face and you know you've done all you can.

We awaited a group photo with Hillary, one of those incestuous campaign traditions that nobody wanted to miss. The group text among the Travelers late the night before went like this:

"Did we get a call time?"

"Not yet, but I heard 9, 9:30."

"Thanks. I don't want to miss the photo!"

"History!"

"Yes. Let's hope she's nice to us."

For nineteen months, Hillary had tried her best to pretend a small army of print, TV, and wire reporters weren't trailing her every move, but that morning she looked tickled to see us.

"Look at the big plane and the big press!" Hillary said, speaking in a baby voice as she stepped out of her black van the morning before Election Day. She was FaceTiming with her granddaughter, Charlotte, and turned her iPhone toward the Travelers as we all arranged ourselves by height in front of the Stronger Together plane.

"Wow! Look at this. Everybody is here," Hillary said, as if we'd be anywhere else.

She spread her arms wide as if she might even embrace the entire mob. She did not. Barb, the campaign photographer, stood on a stepladder. I sat cross-legged on the far left-hand side, the same position I'd assumed on the final day of the 2008 election, when Barack Obama leapt into the middle of his traveling press corps and said, flashing his signature grin, "Say tequila!"

Barb instructed us all to scoot a little to the left or right and take off our sunglasses. The shutter had hardly fluttered when the mob disassembled and crushed Hillary with questions, rendering her a tiny red line in the middle of a voracious scrum. Surveying the scene, the most genial of The Guys, a preppy brunet with a student-body president's grin who traveled everywhere Hillary went and who wore brown oxford loafers even in a New Hampshire blizzard, shook his head. "This is why we can't have nice things," he said.

"You've been often ahead of your time," said a BBC correspondent, pushing her slender mic and soft question in Hillary's face. "You've been sometimes misunderstood. You've fought off a lot of prejudice. Do you think that today America understands you and is ready to accept you?"

Hillary wasn't about to fuck up hours before the polls opened by talking about sexism and her weird, complicated place in history. "Look, I think I have some work to do to bring the country together. As I've been saying in these speeches in the last few days, I really do want to be the president for everybody."

Right before takeoff, an editor in New York called to check in, asking the question *Times* editors stuck in the newsroom always asked— "What's the mood like there?"

"Hillary is orgasmically happy," I said.

I regretted using such a sexual term to describe the woman who, in a matter of hours, would become the FWP, but I couldn't describe her any other way. Through two presidential campaigns, I'd watched Hillary wear her disgust with the whole process—with us, with her campaign, with losing—on her face. The previous summer, I had posted a photo on social media of Hillary at a house party greeting supporters in Ottumwa, Iowa. Within seconds, someone commented, "She looks like she'd rather be at the dentist."

But now Hillary's expression said it had all been worth it. She wasn't just about to become president. Hillary, who until Trump came along had been the most divisive figure in American politics for a couple of decades, was about to become the Great Unifier, relegating Trump and his bullying to the annals of reality TV. Her campaign aides in Brooklyn, all the data, and the early-vote returns assured her he couldn't win.

"We think we're going to do better in the Philly suburbs than any Democrat has in decades," Robby Mook, Hillary's chipper campaign manager, told us. "If we win Pennsylvania and Florida, he just has no path." In other words, *it's over.*

At the election-eve rally in Philadelphia with Bruce Springsteen, Hillary joined Obama onstage. He crouched down a little to kick a step stool closer to her podium. "When you're president, it's gonna be permanently there for you," Obama whispered in her ear before kissing her cheek and exiting stage right.

Later that night, when we boarded the S.T. Express in Philly to fly to Raleigh-Durham for a final "Get Out the Vote" rally with Lady Gaga, the Travelers rushed to the front of the press cabin. We formed a human pyramid in the narrow opening where those of us who didn't mind squatting on our knees and getting crushed by reporter limbs and camera lenses and dangling furry boom mics got a clear view of Bill and Hillary. They were cuddling.

The cynics will roll their eyes at this, but they weren't there. Bill cupped Hillary's shoulder with one of his long hands. He pulled her in tight, under his arm and into his chest, and not in the phony forced way political partners embrace for the cameras. That night, Bill looked at Hillary like she was the prom date he'd wooed all semester. He looked at her like she was the president.

Hillary squeezed him back with a look not of adoration but more like that of a mother trying to control a problem child. Bill glimpsed the press piled up, like coiled springs waiting to pounce. Seeing me scrunched in the bottom front, he said, "Oh, hi, Amy." (Unlike Hillary, who had a gift for looking straight through me as if I were a piece of furniture, Bill always said hello.)

Asked about the significance of the evening, he said, "To finish here tonight I felt was important because that is where the country began."

Then Bill Clinton did what he always did. He made the biggest night in Hillary's life about himself. "It was interesting. You know, I sit on the board of the National Constitution Center . . ."

At that point, Hillary thrust her entire body toward the cockpit, the opposite direction from our scrum, dragging Bill, whose arm was still affixed tightly over her shoulder, With Her.

"Did he just say he was on the board of the National Constitution Center?" a wire reporter, to my right, asked.

"Yes, yes, he did," I said.

"Classic."

ONLY A HANDFUL of Travelers (the "tight pool" in Trailese), including the Wires, a print reporter from one local and one national newspaper, and a rotating TV crew that shared its footage with the rest of the pack, could fit inside the elementary school's auditorium to watch Hillary vote the next morning. I'd spent the past week pleading with The Guys to let me be in the pool on Election Day. In 2008, by a stroke of dumb luck, I'd wound up in the pool in Chicago. I still have my notes: "7:36 a.m., Beulah Shoesmith Elementary School on Chicago's South Side. Obama votes, Sasha and Malia with him."

That night, I'd waited outside the Hilton as Obama and his family and closest aides watched the returns come in. I remember the corrugated metal arm of a loading dock pulling closed over an armored SUV and, like some magic trick, opening again seconds later with the country's first black president-elect inside. From there we rode in the motorcade where 240,000 people waited in Grant Park.

"You have to let me. The *Times* is the *local* New York newspaper. The hometown paper always gets to be in the pool," I begged one of The Guys, a slick newcomer and hired gun to Hillaryland, whom we thought of as the poor man's Ben Affleck because he could've had Hollywood good looks if he didn't spend most of his time like an overly made-up windup doll dispensing legalese about Hillary's emails on cable news. Hired Gun Guy, who'd come up in New York politics, pointed out all the times he'd tried to get the *Times* to cover some small-ball press conference only to have us push back with "We are a *global* news organization, not some local paper."

But my request worked its way up the ladder at the campaign's Brooklyn headquarters and, figuring I couldn't do too much damage at that point, they agreed.

You'd think that after weaseling my way into a spot as the local pooler, I would've used the opportunity for some grand journalistic purpose. Instead, as the press van took us from the Ritz to the elementary school in Chappaqua, where Hillary would cast her vote, I stared out the windows entering a numb, almost meditative state.

To my right, a BMW pulled out from behind black iron gates that swung open to reveal a long driveway that led to a limestone mansion. To my left, the sun came up over the Hudson and painted the sky with pastel peaches and sherbet oranges against the fall leaves.

In the reflection, I saw dark circles under my eyes and flashed back to a sixth-grade slumber party. We'd been upstairs at my friend Heather's house playing Jenga in a carpeted den when a prissy girl from a private school I'd just met asked me if my dad was a pilot.

"I know another girl who has those black circles under her eyes and her dad is a pilot," she said, as if a parent's sleeplessness could be passed down genetically.

Growing up in south Texas, I can't say I ever envied the people who grew up in places like Chappaqua and Rye and Scarsdale, but that's only because I didn't know this Platonic ideal of suburbia existed until my life became intertwined with Hillary's. I'd never given Westchester much thought until that morning when I realized my early ideas about what adulthood should be had been crafted around the problems I imagined the people who lived here had. Problems rooted in stock prices and boredom and private-school entrance exams, ripped from the pages of my rumpled copy of *Revolutionary Road*—and not the batshit redneck things that happened in my 1970s-era subdivision in San Antonio. It occurred to me that of all the people in black churches and union halls and high school gyms and factory floors all over the country whom I'd talked to and who told Hillary their problems, it was the lucky bastards here, behind the secure gates and neat hedges of Westchester County, who got to pick our presidents.

The Travelers hoisted ourselves up onto the wooden stage of the elementary school, resting our heads on each other's shoulders. On the cinder-block wall, a glittery handmade sign thanked the school's janitorial staff: WE SPARKLE BECAUSE OF ADELINO, ALFREDO, HENRY, MANUEL AND MARIO.

All the Hillary faithful showed up. The ones who couldn't fit inside pressed their bodies and their Patagonia fleeces against metal barricades. They held WE BELIEVE IN YOU and HILLARY FOR CHAPPAQUA signs. There were no "Lock her up!" chants in Chappaqua.

Voters lingered in the auditorium, overcrowding the room and forcing security to form a human walkway around Hillary when she arrived as if she were a heavyweight champion entering an arena. That's when everyone exploded, forming a mosh pit of positivity around her. Fathers hoisted up little girls on their shoulders, including one in a pink puffer coat who was entirely too old for a piggyback ride.

Hillary, looking rested even though she couldn't have slept much longer than we did and no longer wearing the thick glasses she'd had on when she greeted supporters at the White Plains airport at the 4:00 a.m. tarmac meet and greet, slumped over to fill out the New York ballot. She extended an arm and gave a wristy wave.

"It is the most humbling feeling," she told us outside the polling station, a tree so red it looked lit on fire behind her. "So many people are counting on the outcome of this election, what it means for our country."

I asked Hillary if she'd been thinking about her mother, Dorothy Rodham, born into poverty and neglect on the day Congress granted women the right to vote.

"Oh, I did," Hillary said, squinting in the bright Election Day sun.

2

Jill Wants to See You

What gives journalism its authenticity and vitality is the tension between the subject's blind self-absorption and the journalist's skepticism. Journalists who swallow the subject's account whole and publish it are not journalists but publicists.

—Janet Malcolm, *The Journalist and the Murderer*

NEW YORK CITY, JULY 2013

I reclined on the exam table. My heels rested in the cold metal stirrups when Dr. Rosenbaum asked me (again) about children. This should have been the start of a heartfelt discussion about motherhood and how to start tracking my menstruation cycle, but all I could think about was Hillary and the election cycle. I did the math in my head. It was 2013. I was thirty-four. Three years until Election Day.

I peered over the tent my medical gown had formed as it tugged tight around my bent knees. The paper crinkled beneath me as I wiggled upright.

"So, how much would it cost to freeze my eggs until after the election?" I asked.

FOUR MONTHS EARLIER, I'd come back to my cubicle at the *Times* to find a sticky note affixed to my desktop. "Jill came by. Wants to see you," it read.

My stomach sank. The air was sticky and Midtown had started to empty out by noon ahead of the Fourth of July weekend. I'd been at Bryant Park eating a salad chopped so thoroughly it might as well have been pureed.

I was wearing a pair of torn Levi's at least a decade old with scraggy seams and holes so wide my knees jutted out. When you reach a certain stature at the *Times*, you can dress like the Unabomber, but I was a media reporter who'd been at the paper less than two years. I couldn't meet with the boss in those jeans. I sprinted through Times Square, past the throngs of tourists and Elmo characters, to the Gap to buy a pair of white pants. They were high-waisted and fell a couple of inches too short around my ankles, but they were on sale, and I could keep the tags on and return them at the end of the day.

I peeked my head in the corner office. Jill Abramson, the executive editor of the *New York Times*, sat on a love seat in front of a wall of windows looking out on Forty-First Street. Her bangs flopped on her forehead and the afternoon light formed a sort of halo around her petite frame.

For me, Jill had been like some very intimidating guardian angel of journalism. Eighteen months earlier, she'd plucked me out of relative obscurity as a features writer at the *Wall Street Journal* to cover media companies at the *Times*. Now Jill told me she remembered reading my Hillary stories in the *Journal*, where I'd covered her doomed 2008 primary campaign before switching over to cover Barack Obama.

2008 seemed like another life. I was twenty-eight and unmarried then, still trying on various personalities to see what fit. I'd already tried Poet, hooking up with men I'd meet at open-mic nights. And Magazine Writer, hopping between assistant jobs hoping that organizing the fashion closet at *Mademoiselle* would somehow lead to a staff writer position at the *New Yorker*. More recently, I'd tried Foreign Correspondent in Tokyo. This included a hot-pink cell phone and regularly spending nights in a jasmine-scented capsule at a spa in Shibuya. In 2007, I experienced the culture shock of going straight from Japan to Iowa to cover the presidential election for the *Wall Street Journal*. Four years later, Jill brought me to the *New York Times*.

I adored the *Times* more than I ever thought it possible to love an employer. Worshipped the place entirely out of proportion. Each time I'd walk in the headquarters, usually stopping to talk to David Carr, the media columnist who was almost perpetually outside smoking, I felt a surge of gratitude mixed with suspicion that someone would figure out that I didn't belong there.

David had survived Hodgkin's lymphoma, and his gaunt frame, gravelly voice, and spindly neck cut a frightening figure for the people he covered. But to me, he resembled a lovable tortoise in a black overcoat, feet up, extending his nape over his cubicle wall, or slurping up a bowl of ramen at his favorite Japanese joint on Ninth Avenue. He may have had to bolt out of the newsroom to meet Ethan Hawke for lunch on the rooftop of the Soho House, but he never lost a mix of folksy Minnesota nice and edginess that reminded me of the people I grew up with in Texas—salt of the earth and sweet as pie until you cross us. He'd wrestled with addiction and mostly worked at alt-weeklies before he landed at the *Times*. He liked that I was from south Texas and that in college I'd worked at a snow cone stand and flipped tortillas at a Tex-Mex restaurant.

One night, David and I were locked in a conference room eating the last of the stale donut holes he'd picked up that morning and trying to chase down a tip about an unscrupulous consortium of New Jersey Democrats and businessmen trying to buy the *Philadelphia Inquirer*. We hammered the publisher and CEO on speakerphone until I finally got him to break down and admit to meddling in the news coverage. David and I silently high-fived each other. After that, David called me the Polar Bear because, he said, "you look sweet and cuddly, but really you're a fucking killer."

In my first years at the *Times*, I spent weeks in London covering the phone-hacking scandal at Rupert Murdoch's British tabloids. And I got to tour the Paramount lot in Los Angeles with Sumner Redstone and a woman in six-inch Lucite stilettos with ample silicone breasts, who his corporate PR team told me was the pervy billionaire's "home health aide." But I missed politics and more specifically, I missed covering Hillary.

On the side, I kept a hand in Clinton coverage during the State Department years. In 2011, I got the first-ever official interview with Chelsea, which doesn't seem like much of a feat now but in those days she told a nine-year-old "kid reporter" with *Scholastic News* that she didn't talk to reporters, "even though I think you're cute." The following year, I joined Bill, Chelsea, and a chartered Sun Country jet full of donors on a Clinton Foundation trip to several African nations. It was late one night at the hotel bar in Johannesburg when Bill told me his daughter is "a very unusual person."

That she was. A couple of nights later, over a South African chardonnay at the Serena Hotel in Kampala, I suggested to Chelsea that we check out the market in the morning. "It's supposed to be the biggest market in East Africa," I said. "Actually, in terms of square footage, Nairobi would dispute that," Chelsea replied.

JILL HAD TATTOOS of a New York subway token and the Old English *T* for the *Times*. She was a stone-hard badass who cut her teeth covering politics and had known Hillary since she was a lawyer at the Rose Law Firm in Arkansas. Jill had been among the post–civil rights movement wave of Harvard-educated New Yorkers drawn to the South. She had more history with the Clintons than most journalists and more foresight than anyone about what Hillary would do next.

"It's obvious she's going to run again," Jill said to me in her unhurried way. "We need you to cover her full time."

I said yes before she even finished speaking. Hillary and Jill, two women at the vanguard and me in the middle.

"I would love that," I said. "Ever since '08 that's been my dream job. I'm so honored you thought of me for this. Thank you so, so much." And then I asked, "When would I start?" thinking Jill would suggest the fall or maybe early next year or after the midterm elections.

She looked at me instead as if I were a small child. "Immediately," she said.

It was 649 days before Hillary would announce she was running for president again, 1,226 days before she would lose to Donald Trump.

IT TOOK YEARS for me to understand the significance of Jill's decision and my own naïveté about what I was stepping into. At first, I embraced my new beat with unfettered enthusiasm; I would be covering the FWP for the paper of record. I considered several of The Guys, especially the originals who'd been with the Clintons for years, friends. I knew about their hookups. I knew which reporters they liked and which ones they hated. I'd met their dogs, rescue mutts. We'd banter about the *Times* staff, and I'd pass on my palace intrigue in exchange for theirs. They'd complain that Chelsea had become a real pain in the ass ("raised by wolves," was how one of them put it), and I'd commiserate with them about colleagues. I even invited two of The Guys to my wedding.

The first of The Guys I called to tell about my promotion to the politics team, I'd known since we met on a frozen tarmac in Elkader, Iowa, in 2007. We'd bonded over a shared love of Jason Isbell and our self-proclaimed outsider status. Neither of us lived in Washington or had any desire to. Of all The Guys, Outsider Guy was the one who I thought transcended the source-reporter relationship, and over the next few years he would become the cruelest, the one whose name I most feared seeing in my inbox. I would eventually create a special DICKHEAD file for his emails. I'm certain that I let him down, too, and that my emails likely wound up in a SNAKY BITCH WHO PRETENDED TO BE MY FRIEND file.

"How cool is that? We'll get to work together all the time," I said.

The line went silent. Outsider Guy's demeanor was as icy as that tarmac had been, and in an instant I knew that we'd never go back to being friends. I thought I heard his pit bull mix growling in the background. The rest of The Guys' reactions continued like that, ranging from stunned ("Uh, okay. You know she's a private citizen, right?"), to aggressive ("Just know you're gonna have a target on your back."), to personal ("You don't get it, do you? Jill hates Hillary.").

The *Times'* public editor, Margaret Sullivan, wrote that the paper's "treatment of Mrs. Clinton as an undeclared, free-agent front runner helps her." Hillary didn't see it that way. The Guys let me know that their hostility came directly from Hillary. She was outraged. She'd hoped to ride the years between the State Department and her next campaign outside the media's glare.

The *Times'* decision to put me on the beat so early fundamentally changed how Hillary's fledgling campaign was covered. Pretty soon, a super PAC called Ready for Hillary gained traction to support her 2016 run. The group became, as one source said, "a make-work program" for old Clinton hands angling to get back in the game. Other news outlets soon announced their own Hillary beat reporters, mostly women: Brianna Keilar (CNN), Maggie Haberman (*Politico*), Ruby Cramer (*Buzz-Feed*), Liz Kreutz (ABC), Monica Alba (NBC), etc. The Hillary press corps had started to take shape three years before the election.

Hillary had a 70 percent approval rating then and hoped to spend her days quietly laying the groundwork for 2016 and her evenings basking in adoration at Manhattan charity galas where she could reconnect with donors. ("Okay, I'm rested!" she'd told a friend when she called before 7:00 a.m. the day after she left the State Department.)

In this period, she'd be feted for saving the whales, combating malaria, working to eradicate adult illiteracy, supporting the Jews, being a Methodist, cracking down on elephant poaching, speaking out against female genital mutilation, rebuilding lower Manhattan after September 11, and popularizing pantsuits.

But it wasn't just that Hillary didn't want media scrutiny. It was something specific to the *Times*. Something larger than me. Bill and Hillary both believed that the paper was out to get them. That may sound irrational to people who think, *The liberal* New York Times, *out to get Hillary?* But they had their reasons.

Hillary didn't see me as I was—an admirer in a Rent the Runway dress chasing this luminous figure around Manhattan and hoping to prove myself on the biggest opportunity of my career. To her, I was simply the latest pawn in the decades-long war that was the *NYT* vs. HRC.

I knew almost nothing about this battle other than that it started around the time of my bat mitzvah. In 1992, the *Times*' investigative reporter Jeff Gerth broke the story about an Ozark land deal gone awry. The Clintons lost money on the development along the White River, but the subsequent investigation into Whitewater would dog the Clinton administration and ultimately lead to impeachment. The thinking went that Howell Raines, the *Times*' Alabama-born Washington bureau chief in the early 1990s, wanted to take down Bill Clinton over some deep-rooted Southern white man rivalry.

I first read about this feud in journals kept by Hillary's closest confidant, Diane Blair. Throughout the White House years, Hillary turned to Diane, whom she'd been inseparable from ever since 1974 when they found each other—kindred, outsiders—in Fayetteville, Arkansas. Diane took detailed notes on their conversations ("Talked books," "Talked about how should she deal with all this shit," "Told her about our cerulean sky") in case Hillary, then the first lady, ever wanted to write a memoir. But when Diane died of cancer in 2000 at the age of sixty-one, her husband, Jim Blair, donated his wife's piles of papers to the University of Arkansas, where Diane had taught political science, having no idea the boxes included some of Hillary's most intimate confessions. I learned about this trove in early 2014 and have pored over its contents ever since.

"She and Bill triumphed despite the press, it heightened their antagonism," Diane wrote in a 1999 entry. "But still, what do you do? Howell Raines of NYT Editor viscerally hates them; wants to destroy."

The relationship with the *Times* went downhill from there.

In 2007, Hillary blamed the *Times* for propping up Obama. A front-page story about his basketball pickup games sent Hillary into a particular rage. "She doesn't have any camera-ready hobbies," the 2008 Guys had protested.

I envied Patrick Healy, the Hillary beat reporter for the *Times* during the 2008 primary. From my perch as a *Journal* reporter, I thought the campaign treated Pat like royalty, always bestowing on him the aisle seat on Hill Force One, always calling on him second at press conferences, after the Associated Press. I dreamed about one day having that aisle seat, getting that second question. But it had all been smoke and mirrors. The 2008 Guys, most of whom didn't stick around for 2016, tried to ruin Pat's life, just like the mix of old and new Guys were gearing up to ruin mine.

In fairness, the torture worked both ways.

The Guys would tell you that I was the worst kind of reporter. Sneaky, a traitor whom they'd given the benefit of the doubt to and who had repeatedly screwed them over in return. They'd say I gravitated to salacious details and always played the victim ("the shrinking violet act," they called it) when all the while I was the one standing over the barrel of ink. I knew they wanted me to be more transparent and honest about what I was working on, but when I'd tried that, it hadn't gone particularly well.

I told them about a feature I wanted to write on how Bill Clinton had taken on a larger role in combating climate change, essentially co-opting the environmental movement from Al Gore, who'd become something of a liberal tree-hugging cliché then. My editors wanted it for page one. Before I knew it, The Guys scheduled a special Clinton Foundation panel in New York. Clinton and Gore sat onstage together in a ballroom at the Sheraton Hotel to discuss working together to combat global warming. Charlie Rose moderated. "We do talk a lot, about everything, but especially about all this energy business," Clinton said.

CBS News called the discussion a "high-profile reunion." But I suspected The Guys had thrown the panel together solely to kill my story. And it worked. I never said they weren't good.

Hillary, meanwhile, was such an avid *Times* reader that over the next couple of years I'd hear that she'd complained about a story's placement in the print newspaper. "Why wasn't it above the fold?" or "Did we get two columns?" The Guys informed me she'd been enraged when she saw that my story about the debut rally of her 2016 campaign, a logistical feat in the middle of the East River on Roosevelt Island, ended up on page A24. I explained that I preferred the front page, too, but the rally had been so late that we'd missed the page-one deadlines. "And almost every other paper in America managed," one of The Guys replied.

The Clintons theorized that Jill Abramson, the first female executive editor of the *Times*, had a personal vendetta against Hillary, something about them both being powerful women at the top of their fields. This "Jill vs. Hill" rivalry was fiction. I saw how much Jill respected Hillary, always had, but she also loved a good story.

This primal instinct to tell a Good Story, the story that people read and share and talk about breathlessly on cable TV, goes back to the dawn of man and always requires tension. The charcoal scrawls of the Stone Age rarely portrayed human-interest stories. The ancient Greeks didn't do puff pieces. Tension means the subjects of the Good Story (in my case the Clintons) often don't think it's good. They think it's a heaping pile of bias ordered up by compromised, click-obsessed editors and written by unscrupulous reporters with below-average IQs. They think it's Fake News from the Failing *New York Times*.

If I wanted to thrive on the politics desk, I would need to do more than feel-good pieces like the ones I'd written on Bill Clinton's charitable work in Africa and on Chelsea taking on a more public role as an NBC News special correspondent. I would need Tension. "You've gotta break some eggs to make an omelet," David Carr would remind me.

MY FIRST FRONT-PAGE story on the beat was about Hillary giving paid speeches for $200,000 a pop to the scrap-metal-recycling expo and the National Automobile Dealers Association in which she offered Mitch Albom–style wisdom. ("Leadership is a team sport." "You can't win if you don't show up." "A whisper can be louder than a shout.") My second was an investigation, cowritten with my colleague Nicholas Confessore, about mismanagement and dysfunction at the Clinton Foundation.

When Dennis Cheng—the foundation's top fund-raiser, whom I got to know on the Africa trip—heard from a donor that I was working on the story, he supposedly said, "Amy? But I thought she was our friend."

Another source likened the Clinton Foundation story to punching the biggest, baddest motherfucker in the prison yard in the face on my first day of a four-year sentence.

"At least they know who you are now," he said.

"Yes, and they could also shiv me in the shower."

Carolyn Ryan, the paper's politics editor and my new no-bullshit boss, made a name for herself at the *Boston Globe* and had New England newsprint in her blood. She'd led the *Times*' metro desk's coverage of New York governor Eliot Spitzer's rendezvous with a call girl, a scandal that ended his career and won her reporters a Pulitzer. Carolyn, who had an infectious guffaw, a mischievous smile, and the spunk required to stroll around the *Times*' newsroom in a Boston Red Sox hat, was such a straight shooter that even after her reporters' coverage led to his ouster, Spitzer sent a video message wishing her good luck on her new job leading national political coverage.

At first, it was just me and her and a handful of political reporters scheming up stories that she would then edit and pass on to the copy desk, a grizzled group of editors who saved us from ourselves, scanning our stories for factual errors and slang that didn't fit the *Times* stylebook. (For years, hardly anyone "tweeted" in the *Times*. They "wrote a message on Twitter." There was no "email," only "e-mail.") Copy editors then passed the story on to the slot, another collection of editors (named after the old days when newsprint would be whizzed through a slot to the printing press). The slot editor would give the story a final read before sending it into the abyss until it arrived on doorsteps the next morning.

But the seedlings of the story always began with a reporter and editor talking. Carolyn had a more innate sense of what people wanted to read and a more natural ability to get the best out of her reporters than any editor I'd ever worked for. Talking to her set every brainstorming session off on rollicking tangents that included gossip collected in the congressional dining room, on the Washington softball field, and while waiting for the *Times*' vending machine to spit out some stale Twizzlers. Unsubstantiated tidbits—particularly involving Bill and Hillary, Elizabeth Warren, and anything related to New York politics—would cause Carolyn to leap across her desk with a "No way!" and "We gotta get that in the paper." And once the first draft was written, Carolyn's editing style was like an episode of *Antiques Roadshow*. In minutes, she could squint her sea-blue eyes at the screen, sweep her yellowy bangs out of her face, and weed through two thousand words of crap, pulling out a priceless treasure of an anecdote buried in graph fifteen.

The Guys hated the kind of memorable details that Carolyn and I both gravitated to. They could forgive us for writing about potential conflicts of interest at the Clinton Foundation and Teneo Holdings, the shadowy advisory firm cofounded by Doug Band. Doug had a thinning hairline that made him look both older and more dignified than his forty-one years. He'd meet *Times* reporters for lunch at Il Mulino and slap twenty-dollar bills into the host's palm, a practice I'd only ever seen in Mexico City. Doug started in the White House as an intern and became Bill Clinton's closest aide in the post–White House years, parlaying his role into a profitable private-sector gig. One of the '08 Guys used a *Downton Abbey* reference to sum up Doug's position in the House of Clinton: "Doug forgot that he lives downstairs." The Guys welcomed negative stories about Doug. He was the perfect scapegoat for all Bill's questionable behavior, as if the forty-second president were just a lovable St. Bernard. *He wouldn't care about making money and about swanky hotels if it weren't for that Doug Band guy . . .* The most astute mind in American politics reduced, in their spin, to slobbery obedience.

But they could never forgive me for the Yorkie.

I had a detail about the foundation purchasing a first-class ticket for Natalie Portman and her beloved dog to fly to one of its Clinton Global Initiative gatherings. Carolyn loved the Yorkie. She wanted to make it the lead.

"It's a fucking Yorkie, Amy!" Outsider Guy yelled as I stuttered trying to explain why this was a critical detail that showed the charity's glitzy overspending. "It weighs like four fucking pounds. It's not like it needed its own seat on the plane."

A year later a conservative super PAC sent around an anti-Hillary fundraising plea: "The Clinton Foundation—which pays to fly her around on private jets, flew Natalie Portman's Yorkie first class."

Carolyn emailed me, "I knew that Yorkie would be back."

I STOPPED THINKING of The Guys as individuals. There would be departures and firings and new hires in the Clinton press shop, but they were all the same to me, a tragicomic Greek chorus hell-bent on protecting Hillary and destroying me. "You've got a target on your back," The Guys always told me, like the drumbeat of failure foretold.

They called the *Times'* politics team a "steel cage match." They'd feign concern: "I just don't want to see you become the Jeff Gerth of your generation." Whatever that meant.

They told me that Jonathan Martin, a sweetheart of a colleague who had recently joined the *Times* from *Politico*, was telling Maggie Haberman, his former *Politico* colleague who scooped me daily, what I was working on. (He wasn't.) I later accused The Guys of telling Maggie what I had in the works. This not only wasn't true and made me sound like a psychotic ex-girlfriend, but the accusation marked the beginning of the end of any semblance of a cordial working relationship. They'd gotten in my head, and I let them. I believed The Guys when they'd warn me that my more assertive (male) colleagues would boot me off the beat and tramp over my bloodied corpse.

"In one corner of the steel cage match, Healy and Hernandez. In the other Chozick . . ." The Guys said. (Healy and Ray Hernandez would become close friends, it turned out.)

But mostly, The Guys loved to say, "I saw what happened to Anne . . ."

They meant Anne Kornblut. Jill Abramson put Anne on the Hillary beat during the Senate years as Hillary prepared to run for president in 2008. The way The Guys told it, Anne had been done in by *Times* political reporters, with Pat and Ray leading the effort to oust her from the most sought-after beat in journalism. In reality, Anne left the *Times* in 2007 for a generous offer from the *Washington Post*, where she won a

Pulitzer and ended up practically running the place before leaving for a high-powered job at Facebook. Anne turned out just *fine*.

I felt like a hazed fraternity pledge, aware that even as The Guys tormented me, I needed them.

Not long after Sheryl Sandberg presented Hillary with the Women for Women International Champion of Peace Award, I was on a conference call with five of The Guys, going through a list of facts to check for an upcoming story. I tried to negotiate the use of some innocuous color I'd gathered over a casual meal with the high chieftain of The Guys, the longest-serving Svengali and the most-devoted member of Hillary's court of flattering men. He was the OG, the Original Guy.

"Absolutely not," OG said.

I groveled. "But you didn't say it was off the record."

"I didn't know I had to say it was off the record when I was inside you," he replied, paraphrasing a line from the movie *Thank You for Smoking* in which Aaron Eckhart plays a slick tobacco flack who is sleeping with a plucky young reporter played by Katie Holmes.

Inside you.

The words hung there so grossly gynecological. On the upside, at least I was Katie Holmes in this scenario.

I started to feel alienated in the newsroom, paranoid about whom I could trust. I stopped having lunch in the Times cafeteria. I even missed burrito day, and I lived for burrito day.

The Guys would time their rants (subject line: "HRC/NYT") to land in my inbox on Friday nights or Saturday mornings, usually when I was walking into a spin class ready to give myself over to an instructor in a sports bra imparting self-help wisdom. ("Who you are in this room is who you are in life.") But all I heard over Drake was *You've got a target on your back.*

I still felt some kind of a feminine bond with Hillary then, even though in the months I'd been on the beat, I'd only talked to her for two minutes in a freight elevator in San Francisco after I followed her out of an American Bar Association conference. I assumed she kept The Guys around because they were entertaining. (When the RNC placed a fuzzy orange squirrel outside a Hillary event with the words ANOTHER CLINTON IN THE WHITE HOUSE IS NUTS, one of The Guys said, "Wait, I think I fucked that squirrel in 2008.") They were handsome (by Washington standards). They had potential.

Maybe Hillary wanted to mold them into better men. After all, I'd spent my midtwenties dating an Italian documentary filmmaker who my friends pointed out was more like a homeless man with a camcorder. Didn't all women have an unspoken urge to nurse damaged men who worshipped us?

But then that was me doing what I so often did—imagining Hillary as I wanted her to be and not as she really was.

Months later, when I explained to my mom why I needed her to violate controlled-substance laws by filling her unused Xanax prescription and FedExing me the pills, she said, "It's such a shame. If only Hillary knew . . ."

It took me years, but when my grasp of the real Hillary finally came into focus, I accepted that it wasn't that she didn't know how The Guys acted. It was that she liked them that way.

BY THE TIME Dr. Rosenbaum puckered her severe features at me and snapped off her rubber gloves, I'd already picked up my phone and was scrolling through Twitter. She suggested I get pregnant as soon as possible. "Take an au pair on the campaign trail," she said. "I have a lovely student from France. The twins just adore her."

I nodded and imagined piling into a press riser in a high school gym for a campaign rally in Cleveland with an infant, a French au pair, and possibly Bill Clinton nearby.

Until that afternoon, I hadn't grasped how intrinsically linked my own life and Hillary's pursuit of the presidency had become. I threw on my clothes, rushed across town back to the newsroom, and made a mental note to find a new doctor.

It was Hillary Clinton vs. my ovaries.

3

"The World's Saddest Word"

SAN ANTONIO, TEXAS, 1996

Doris and my mom worked together at a public school with half a dozen or so pregnant thirteen-year-olds and a vice principal who'd served in the Marines and once wrestled a gun away from a seventh grader. It was Doris who told my mom about her big plans for that weekend: the first lady was in town promoting her child-rearing book, *It Takes a Village*, and Doris was going to hear her speak.

I can see my mom in the staff room at Pat Neff Middle School, pulling an orange out of the brown paper bag she packed for herself and thinking about this concept of the village raising a child. That same year she would be diagnosed with, and later recover from, breast cancer. I'd bring her strawberry popsicles that stained her lips red during chemo. Most people in Texas adhered to Bob Dole's belief that "it takes a family to raise a child" and saw the whole village thing as a commie concept from a radical, uppity first lady. The *San Antonio Express-News* would describe the women's organization represented at the book event as "'feminists' (whatever that means)."

My mom must have told Doris that I had a budding interest in politics, or maybe Doris offered to take me to meet Hillary Clinton, but either way, Doris picked me up early Saturday morning in her white Cadillac DeVille coated in a layer of saffron-colored pollen. The interior smelled like cigarette smoke, and a dangling air freshener shaped like a cowboy boot hung from the rearview mirror. Doris wore her hair in a dyed-black beehive that practically rubbed against the car's interior roof. She told me she was a psychic and predicted I'd write children's books one day. She asked me about the tennis team. I told her I'd quit. She told me my mom was proud of me. I said nothing.

Doris signed herself in at the Hilton Hotel conference room overlooking the River Walk and grabbed us a bar table by the windows. She brought me a Coke. I looked at my Swatch watch. I had no idea that would be the first of hundreds (thousands?) of times I'd find myself waiting on Hillary. Clinton Time, I'd learn to call it. By the time Hillary arrived that afternoon at the Hilton, I'd been through four Cokes. Doris smiled, her caked-on makeup cracking around her eyes. She pulled my wrist and led me to the front of the room where Hillary took her place behind a microphone.

"Go! Get in there. Get close," Doris said.

I don't remember anything Hillary said that day. But I remember the feeling I had when I saw her, the caffeine and adrenaline, the rush of a real-life celebrity who was not Selena or a member of the Spurs, in my hometown. She was pretty. She wore some version of pink or blush, definitely pastel, and looked like the kind of woman who might have belonged to the Junior League. (I later learned Hillary agreed with Anna Quindlen's characterization of the role of first lady as having to be "June Cleaver on her good days.")

I didn't know then that Hillary hatred was already, as the author Garry Wills called it, "a large-scale psychic phenomenon." Or that the RNC sold Hillary rag dolls that could be dismembered. Don Imus played "That's Why the First Lady Is a Tramp" on his radio show. ("She won't do housework because it makes her sick, doesn't bake cookies like the rest of the chicks . . .") But I knew my friends all hated her, which meant their parents must have hated her, too. I didn't know why. She didn't look scary to me.

I made my way to the front of the hundred or so women and reached my hand out to shake Hillary's. I hadn't thought about what I'd say to the first lady, and all I could spit out was "I'll be old enough to vote in September and I'm going to vote for your husband." I may have let her speech drift in one ear and out the other, but I can hear myself so clearly say those first words I'd ever say to Hillary: *your husband*. Not Bill Clinton, not President Clinton. *Your husband*.

Hillary shook my hand and held on for a while. She leaned down a little to meet eyes with me. She thanked me, and I hear her saying, "It's terrific you're already thinking about voting. We need you!" Then Hillary disappeared out a side door with a couple of Secret Service agents trailing behind.

My mom asked how the afternoon went.

"Fine," I said, pulling ranch dip out of the fridge. "I shook Hillary's hand." Then my seventeen-year-old self said what Hillary the candidate would struggle and ultimately fail to make the country say: "She seemed nice."

That was it. My first astute political assessment of Hillary Clinton. *She seemed nice.*

I AM A fifth-generation Texan Jew, the youngest of two daughters of a public school teacher from San Antonio and a self-employed attorney born and raised in the Baptist heartland of Waco. We were curiosities amid the megachurches and the Hobby Lobby stores and the fast-food restaurants with signs out front that say CLOSED ON SUNDAY FOR FAMILY AND WORSHIP. My friend Jenny gave me a silver cross with a dove in the middle hanging on a delicate chain by James Avery, a Hill Country craftsman who specialized in Christian-themed jewelry . . . for my bat mitzvah.

Politics became inseparable from religion, from our otherness. Jews had big noses and frizzy hair, and everyone assumed, correctly or not, that we were—*gasp*—Democrats.

I might as well have pulled on a skullcap and recited my haftorah when I told Mrs. Shepard's fourth-grade class that I was supporting Dukakis. My parents took me to meet Ann Richards once. I remember her white bouffant and reaching my entire body over a heavy wooden desk to shake her hand. But I couldn't have told you whether my parents were Democrats or Republicans. Politics wasn't something that came up a lot in our house. If presidential politics reached our family at all, it was some homework assignment my dad helped us with or background noise on the TV as my exhausted mom got home from work, threw on jeans and a T-shirt, and tossed into the oven canned crescent rolls and chicken strips.

Yet we couldn't escape local politics.

Sometime in the 1990s, the Texas legislature decided that public school kids, in addition to reciting the Pledge of Allegiance to both the American *and* Texas flags, should also begin each school day with "one minute of silence." Everyone knew this meant Jesus. My parents told me to sit down quietly after the pledge and skip what teachers called the silent prayer. I decided to boycott the morning ritual altogether.

Mid-prayer, Mr. Mack, a photography teacher and Vietnam vet, cracked one eye open, noticed me sitting down, and instructed me to "stand the hell up." When I shook my head no, he kicked me out and gave me three days' detention. I was shoving my notebook and Epson Luster paper into my JanSport with a lot of eye rolling and zero sense of urgency when a linebacker who sat across from me gave us all a civics lesson. "We're a *Christian* country," he said. "It's called one nation under GOD."

BY THEN I'D grown out of what my sister Stefani called my giant dork stage, when I wore tortoiseshell glasses and had my head buried in books, Jack Kerouac and Oscar Wilde, years before I really understood them. I even saw myself in Chelsea then. We were about the same age, from neighboring Southern states, both avid readers and uncomfortable in our skin, with smiles full of braces, curls we couldn't control, and frilly dresses with bubbly shoulder pads. I then graduated to my jock stage when I played varsity tennis and was a starting point guard with a reputation for excessive personal fouls. By the time I met Hillary, I was well into my stoner poet stage, during which I maintained an A average while spending most of my junior year in the parking lot of Rome's Pizza hotboxing my friend Kate's cherry-red VW Beetle while reciting Nikki Giovanni poetry.

Years later, when I came across Hillary's college letters to her own high school friend, I thought of these stages and our shared adolescent misanthropy. "Can you be a misanthrope and still love or enjoy some individuals?" Hillary wrote to John Peavoy when she was a sophomore at Wellesley in 1967. "How about a compassionate misanthrope?"

She wrote about the "opaque reality" of her own self and confessed "since Xmas vacation, I've gone through three-and-a-half metamorphoses and am beginning to feel as though there is a smorgasbord of personalities spread before me," including "alienated academic, involved pseudo-hippie, educational and social reformer and one-half of withdrawn simplicity."

I didn't care enough about anything to belong to the Young Democrats (if such a thing existed in my public school, I wasn't aware) or the debate team. I didn't pass around petitions to end the death penalty and didn't have much of an opinion about the news of the day, even though my dad was from Waco and everyone wanted to ask me about the Branch Davidians and if I knew David Koresh. "That was *outside* of Waco," I'd say.

I decided, for no other reason than that it would piss off every football player who stood around the kegs of Shiner we'd set up in the middle of a field on Saturday nights, that I hated the Cowboys. I didn't eat red meat. I couldn't wait to get the hell out of Texas and move to New York. I loved Bill Clinton . . . and, worst of all, I loved his wife.

I THOUGHT THINGS would be different in Austin. I didn't need ivy affixed to a sandstone library, and as my dad reminds us whenever Stef and I bemoan that we never really had a chance to go anywhere besides the University of Texas, "It wasn't like Harvard was knocking our door down." But I envisioned something artsy—conversations about Camus over absinthe, maybe—something more than dope bud, a Ben Harper show, and seven of us splitting the same bowl of queso at Magnolia Cafe.

I had even more disdain for the sorority girls, the "debutantes," than I did the druggies "expanding" their consciousness, as Hillary summed up both social castes in her college letters. I counted down the days until I could move to New York and become a writer.

My closest friend, Barry Dale—who theorizes that we'd found each other in middle school and both wanted to move to New York because "I was the gay and you were the Jew"—had an assignment in his film class. He needed a model to sit in an empty diner in downtown Austin to re-create Edward Hopper's *Nighthawks*.

I wore a tight black tube dress with a deep V-neck from Bebe, fishnet hose, and a pair of shiny black heels I'd bought a couple of years earlier with my employee discount at Banana Republic. I sat cross-armed at a bar table as Barry stood on the sidewalk outside snapping photos through the glass.

That's the photo I think of from my college years. Not drunken spring break nights or fraternity toga parties or eating stale pizza on deadline in the basement of the *Daily Texan*. Me, in an almost deserted diner, wearing black, looking slightly slutty in a mall-bought dress, staring forward and down at nothing and everything. Barry got an A+. "I love the feel of the girl," his professor wrote of the photo.

It should've been titled the same woe-is-me sign-off that Hillary used to close her college letters.

"Me (the world's saddest word)," she wrote.

4

Bill Clinton Kaligani

SOUTH AFRICA, 2012
Bill Clinton was holding a glass of chardonnay but not drinking it the night I walked into his suite at the Saxon Hotel in Johannesburg. It was after midnight. I'd just flown to South Africa in a cramped coach cabin with a team of teenage rugby players who were bursting with testosterone and fist pumping during the entire sixteen-hour flight, plus a refueling stop in Senegal.

I'd checked into my room in the main house of the Saxon, once the palatial private residence of Douw Steyn, an eccentric billionaire who befriended Clinton during his presidency. As I walked toward Clinton's private luxury villa, I passed rows of photos of Steyn with a younger, plumper Clinton. I crossed a wooden bridge over a pond, the sound of peacocks and fireflies and the hum of cicadas in the distance. I opened the villa's heavy engraved doors. The stand-alone suite had a private bar and a living room decorated in tasteful neutral hues with a scattering of African sculptures.

A handful of Friends of Bill, also known as FOBs, sat at a nearby table playing oh hell!, Clinton's card game of choice. They made small talk about Hillary's 2016 prospects. ("If Romney wins, the party will have to pave the way for her . . .")

Clinton stood by a row of neatly arranged beige leather bar stools, wearing a baby-blue V-neck cashmere sweater, tan driving shoes, jeans, and a friendship bracelet tied around his frail wrist. Chelsea sat on a sunken taupe sofa sipping Evian alongside Bari Lurie, her chief of staff.

I'd later confess to one of the donors, Raj Fernando, an algorithmic trader in Chicago, that I felt guilty about how much the *Times* had paid to send me on the Clinton Foundation trip—a six-night swing through Mozambique, South Africa, Uganda, Rwanda, plus a pit stop in Cyprus so Clinton could deliver a paid speech. "Believe me," Raj said. "I paid more."

It was the summer of 2012, right before Clinton's spellbinding speech renominating Obama at the Democratic National Convention, when no one was paying much attention to Bill Clinton. I'd been in Sun Valley, Idaho, chasing down media moguls and crashing a cocktail party with Wendi Murdoch ("Rupert hates the *New York Times*, but I love you!") when Jill Abramson approved the Africa trip, never mind that it had nothing to do with my beat at the time.

Looking back, it's astonishing that The Guys ever allowed me to cover this philanthropic swing. We were all so simpatico then that when the *Times* photographer showed up from a stint in Yemen with no luggage, Clinton loaned him a razor. I was the only reporter who stayed the entire trip, starting with that first night at the Saxon when Bill Clinton talked my ear off well into the early-morning hours.

Among about a million other topics, he explained that Nelson Mandela had written his memoir on the grounds of the Steyn mansion before it became a five-star hotel.

"Where you're staying was his home, and that's where I stayed until 2010," Clinton said. He looked around at the villa, with its high, airy ceilings and spotless marble floors. "It's a wonderful place. I love this place," Clinton said.

He paused for a moment. He'd visited the Soweto slums earlier that day. The next day we would fly to Rwanda where we'd take a military helicopter to a red-dirt village to visit a children's hospital.

"Yeah, I always feel slightly guilty staying here," he said. He took a sip of chardonnay. "But I get over it."

OVER THE NEXT six days, I vacillated between awe at Clinton's brainpower and verve—feeling blessed to be in this brilliant man's presence—

and total exhaustion from his self-absorption and driveling on. After a couple of nights of hotel-bar banter, I began to feel like the lucky passenger upgraded to first class on a transatlantic flight only to wind up next to a raconteur who never needs a nap and who rambles on because of some internal hole they need to fill.

By the final night in the Kampala Serena Hotel in Uganda, Clinton has relayed his own obscure accomplishments ("In Arkansas, we went from forty-eight percent to fifty-three percent forested land when I left office . . ."). He has summed up how to solve Africa's food shortage ("We need to do things Americans did literally eighty years ago during the Depression . . ."), and he has, for what feels like hours, extolled the virtues of soybeans ("You can grow it with just a thin layer of topsoil . . ."). He starts every other sentence with "In the 1990s . . ." and "When I was president . . ." The Guys even had a name for one defensive monologue I got trapped in after asking about Clinton's decision to invade Somalia in 1993. "You got Black Hawked," they said.

It was after 1:00 a.m., and all I wanted to do was go to sleep when Clinton told me his advice for how Obama could improve his speech making. "Suppose we've been friends for forty years," he said, resting a palm on my shoulder. "If you came to visit me in the hospital and said something pretty and eloquent instead of saying, 'God, I'm sorry. This sucks. I wish I could do more about it,' it's an insult. So I told the president the eloquence should go at the end of his speeches now, never in the middle . . ."

I nodded, smiling politely and checking that the red light of my voice recorder was still glowing.

He changed outfits at least three times a day, usually reappearing in the verdant hotel gardens for dinner with donors wearing a linen guayabera and khaki cargo pants. Africa chic. "He's like Lady Gaga," an aide said.

The other thing I noticed about Clinton was how often he talked about dying. He hardly thought he'd live to see the 2016 election, never mind wind up back in the White House.

When the manager of a soybean processing plant asked him to come back next year, Clinton said, "I'm older than you. We have to make sure I'm still around." When I asked him about Chelsea recently joining the Clinton Foundation board, he said, "We're trying to build it up so it'd still run if I drop dead tomorrow."

In Nicosia, we sat down for coffee, and when Cypriots weren't swarming him for photos, I asked whether Hillary would run for president in 2016. "She points out that we're not kids anymore and a lot of people want to be president," he responded.

I saw things in Africa that made me less cynical about the Clinton Foundation. Under tamarind and mahogany trees, aid workers set up a station where deaf children from the local villages could be fitted with their first hearing aids. It's hard to care about whether some sleazy foreign donor wants something from the State Department after you've seen a child hear for the first time.

And when the Clinton Foundation is maligned, I think of Bill Clinton Kaligani. We were all standing on the tarmac at the Entebbe International Airport, and I'd completely run out of topics to ask Clinton about. I just extended my voice recorder to pick up his stream of consciousness when a military helicopter emerged on the yellow-orange horizon.

"Is that him?" Chelsea said, cupping her hand over her eyes as she looked into the setting sun.

Moments later, a slender fourteen-year-old Ugandan boy in his threadbare school uniform stepped out of the helicopter. His name was Bill Clinton Kaligani. His mother had named him after Clinton when he visited Uganda in 1998.

A photograph hangs in the Clintons' Chappaqua home. Clinton is holding the newborn as Hillary, in a wide-rimmed *Out of Africa* hat, looks on. "He was born the day before we got there," Clinton told me over the hum of the helicopter. "It was one of the most memorable days of my presidency."

He walked over and pulled little Bill into his arms. The boy wrapped his hands around one of Clinton's hands and rested his head on that doughy spot on the chest beneath the shoulder. They stayed there like that.

After they visited for a while, and Clinton said he'd pay the boy's school tuition fees, the staff and donors prepared to board our chartered 737. Aides tugged Clinton toward a separate Gulfstream, but he wasn't done. He called me over and told me that on that same Africa trip in 1998, a Senegalese farmer had named a goat after him.

"We're going to fly the goat in next," he said.

5

Roving

Two weeks after college graduation, I took a one-way flight to New York with no job, no apartment, and a stack of clips from the *Daily Texan*. I'd saved some bat mitzvah money and what I'd made working in a snow cone stand off Barton Springs Road in Austin.

I temped all over Midtown, insurance offices and nonprofits mostly. Before work, I would run around Midtown in my suit and tennis shoes and drop off my stack of clips with the mailrooms or security desks at *Newsweek*, *Time*, *Fortune*, etc. A month later, I got a job as a rover at Condé Nast, the publisher of magazines like *GQ*, *Vogue*, *Vanity Fair*, and the *New Yorker*.

Condé Nast rovers were sort of like temps except that they had six months of steady work. We ranked beneath interns—who almost all had parents on the Upper East Side and attended a Seven Sisters school— but slightly above the outside temping agency I'd been working for. We had half a year to impress editors and publishers, demonstrating our efficiency at cleaning out closets, changing coffee filters, or picking up dry cleaning. Occasionally, we got to answer a phone. If no one wanted to hire us after six months, the gig was up and the program would be replenished with a new crop of desperate, broke aspiring magazine writers.

We all wanted our chance to temp at the *New Yorker* and *Vanity Fair*, but those magazines rarely asked for rovers. Mostly, we helped executives on the terrifying corporate floor, as white and sterile as a hospital, and were rotated in at magazines like *Brides*, *Self*, and *Allure*. One senior editor told me during an interview for a job labeled "beauty closet organizer" that "the best part about working at *Allure* is that it's like a soror-ity." Another asked if I would describe myself as a "makeup junkie," as if one look at my almost-bare face didn't answer that question.

Every day at Condé Nast was an education, though not necessarily the one I thought I'd be getting.

I was opening mail on the corporate floor for a creative director, a towering, hulking bachelor who wore ascots and had me print out all his emails and then transcribe his scribbles in No. 2 pencil in the replies. I got to a package containing a heavy Lucite box full of sand and aqua seashells, an invitation to a party in Montauk with the *Sex and the City* cast. He told me to RSVP no, adding, "But do me a favor. Find out if that actress who plays Charlotte is single." I asked, in the docile voice of domestic help, if I could go to the party in his place.

"How are *you* going to get to Montauk?"

"Uh, well, I'll take the bus." I had no idea that Montauk sat at the tippy-tip end of Long Island.

"You're gonna take the bus? To a Peggy Siegal party?" Ascot said. He laughed, more amused than irritated, and walked back into his office with the cream-colored carpet and the two white leather bucket chairs that I once heard his assistant describe as "the Herman Millers." The furniture had a name, and I didn't.

The one time that I'd rotated into *Vogue*, helping to organize the small closet converted into an office and library for the magazine's fact-checkers, I'd worn my brand-new pair of UGG boots. Anna Wintour walked by the closet, paused, and behind her black glasses, fixated on the furry brown boots that I'd spent half my paycheck on. I never wore the UGGs again.

I'd been crashing at my college boyfriend Russ's place in Bed-Stuy. Back in Austin, Russ and I had bonded over our love of *Midnight Cowboy*, each of us waxing poetic on its artistic subtleties and our own dreams of moving to New York. I should've known the relationship was doomed given that this movie, the ultimate testament to big-city failure, was what brought us together. Russ had an internship at the *Nation* that paid a stipend of $150 a week, and he liked to remind me regularly that

I worked at lightweight capitalist rags and that he was the serious/journalist in the relationship.

Fifteen years later, a Bernie Sanders supporter read my biography online and called me "a gruesome cross between *Midnight Cowboy* and *Working Girl.*" Pretty much. *I'm brand spankin' new in this here town and I was hopin' to get a look at the Statue of Liberty . . .*

MY BOSS AT Condé Nast had me come into the office the morning of September 12, 2001. I walked through a nearly deserted Times Square. I tried to remind this editor that no planes were flying, but she was adamant that she get to a photo shoot in Paris.

On my way to the office, Russ finally called. Other than a brief phone call the day before when we both had to evacuate our Midtown offices, I hadn't heard from him and was starting to worry. He was driving his Honda Civic from its alternate-side parking spot in Fort Greene all the way back to the driveway of his mom's house in Tulsa, presumably listening to Dostoyevsky on tape. He said he was already in Missouri and had decided to move back to Austin. He'd been the only person I really knew in New York, and he'd abandoned me.

In that first year, just existing in New York exhausted me. I always got on an express train when I needed a local, watching fifteen superfluous stops fly by. I couldn't walk three blocks without getting stopped by Greenpeace volunteers or some man with a clipboard who wanted to ask me a question about my hair. Not wanting to be rude, I'd always stop. *No, I don't have a perm. No, does it look like I use a deep conditioner?*

On most nights, I'd collapse, fall asleep fully clothed with the lights on. I'd wake up between the hours of 2:00 a.m. and 6:00 a.m. in a pile of saliva, the sound of sirens outside and the shape of the links from my silver Seiko imprinted on my cheek.

FIVE MONTHS AND three weeks into my rovership, I got my first full-time job in New York. I would be the editorial assistant to the garden editor at *House & Garden* magazine. In the interview, Garden Editor, a fashionably malnourished redhead who wore fishnet stockings and leather skirts, had growled a little when I complimented her leopard-print blouse. "Careful," she said. "It's a jungle out there." Her husband would leave messages like "Tell her I'll be home at six and will need something hot to eat and cold to drink . . ."

My duties included running to the Flower District on Twenty-Eighth Street each morning, where the sidewalks become an urban jungle of houseplants and cut flowers that supply the city's restaurants and hotels and penthouses.

I once got on the Condé Nast elevator hauling cherry blossom branches wrapped in butcher paper when an airy ballerina of a girl about my age strode on, an Hermès Birkin bag slung over her forearm. I only identified this purse—and its $10,000 price tag and waiting list—because of the week I'd spent in the fact-checking closet at *Vogue*.

"Oh my God, I love your bag. Is it new?" another gazelle of a girl asked.

"No," the ballerina replied. "I got it like a week ago."

I tried to time my commute so that I could share the elevator with *New Yorker* editor David Remnick. Stalker-like, I craved even the tiniest reminder (the back of a brown head of hair) that my dream was still in my grasp.

One time, I stepped off on the eighth floor, with the purple *House & Garden* awnings, wearing my usual brown plastic banana clip and thinking about my illustrious future writing gig at the *New Yorker*. When the doors started to close, I overheard a woman announce to the packed elevator, "Okay, who told her she could wear her hair like that?"

Garden Editor would disappear for weeks to scout luscious private grounds in England and France and Morocco. She'd come back with a Longchamp tote stuffed with receipts for me to file and reams of film for me to develop into slides and arrange by theme ("Tuscan," "xeriscape," "Cape Cod"). She'd then present the hundreds of photos, which looked identical to the untrained eye, to the editor in chief, who would deem the gardens sufficiently tony (or not) to earn a spread in the magazine.

One afternoon I was arranging the slides into plastic sleeves, and after fifteen images of topiaries and a couple of bonsai, I landed on photos of the garden editor sprawled out naked, posing seductively for Mr. Hot-to-Eat.

In retrospect, I could've handled the situation more professionally, but I've never claimed to be a good editorial assistant. My gasp must've come out louder than I intended because a crowd of colleagues assembled around my cubicle. They held the slides now scattered over my desk up to the light howling that the garden editor "isn't a real redhead after all."

It was at that exact moment that Editor in Chief walked by and set her Siberian husky–like eyes on me as if I were some game animal she wanted to mount to the wall of her Pelham colonial.

She tried to have me fired for being "indiscreet" and embarrassing the garden editor. Luckily, the managing editor took pity on me (or anticipated a workplace lawsuit) and talked the editor in chief out of it.

I needed the paycheck and the health insurance, but part of me wished she had fired me so I could've filed for unemployment. My roommate at the time had just landed a real writing gig at *Variety* after being laid off from his job at AOL's Moviefone. (He'd imitated the languid voice that reads showtimes: "To file for unemployment, press 1. To clear out your cubicle, press 2 . . .") I envisioned parlaying unemployment into writing an explosive tell-all in *New York* magazine or the *New York Observer* called "The Garden Editor's Bush."

I knew my time at *House & Garden* was up when James Truman, the Condé Nast editorial director who would soon resign to bum around Andalusia and continue his studies with Tibetan Buddhists, flipped through a rough copy of the May issue and instructed Editor in Chief to swap out the cover story for a tiny blurb I wrote on a Chelsea flower shop that stuffed carnations into discarded Café Bustelo cans. This so infuriated Editor in Chief that she sent around a note reminding the entire staff (cc'ing me) that editorial assistants should under *no circumstances* be allowed to write.

I printed out the email and tucked it in my battered copy of *A Confederacy of Dunces*, which sat on my new IKEA bookshelf along with two volumes of poetry by Elizabeth Bishop and Isaac Babel's *Complete Works*, which I'd borrowed from Russ and never given back.

The dunces, all in confederacy against me . . .

I TOOK A pay cut to work for a fancy literary agent who wore black leather pants and was on a strict South Beach diet. She'd email two words, "Protein Run," and off I'd go to buy her hard-boiled eggs or almonds.

Every Friday we had to print out all our email correspondence from the week, and she'd hand the pages back to us on Monday marked up in red ink where she'd fixed typos and stylistic errors. Or, in my case, she explained at length why it was inappropriate to ask the agency's clients for career advice. She had a point, but the only reason that I wanted the job was because her roster of authors included some of my journalistic heroes.

"You can do that. I mean, we all do that, but don't *ever* include it in the correspondence. Duh," another assistant in the Park Avenue office told me. She knew about my screwup because in addition to Literary Agent editing our emails, her five or so other employees all had to read each other's marked-up correspondence. This led to grammatical shaming in the break room.

"Can you believe he used the passive voice in a message to *Knopf*?" this same assistant said as she showed me how to arrange a hamburger patty on a bed of lettuce as Literary Agent liked. "You'd never think he was a Rhodes scholar."

Just when I was starting to appreciate this semantic sadism as a useful crash course in email writing, Literary Agent fired me. She thought I'd stolen office supplies, which wasn't technically true. She asked me to order twelve of her preferred purple highlighters and instead I'd accidentally ordered twelve cases (each of which contained ten highlighters). Paranoid she'd see this Mount Everest of purple in the tiny supply closet and erupt, I took ten of the cases home and put them under the bed figuring I'd gradually restock the office supply with this stash. But as I tried to explain this, Literary Agent just put up her hand in a please-stop-talking position.

Ten years later, I was still pulling purple highlighters from underneath my bed.

I applied for assistant positions at *Cosmo*, the *Economist*, the *Nation*, *Redbook*, *InStyle*, *House Beautiful*, and *Martha Stewart Living*. I interviewed to be Lloyd Grove's research assistant for his gossip column at the *Daily News*. (I heard he preferred young female assistants.) But he took one look at me and could tell I didn't know East Hampton from East Rutherford.

I waited and waited, clutching a faux-leather padfolio filled with clips from the *Daily Texan*, in the lobby of the old Times building. A political reporter whom my sister met through a mutual friend said she'd meet with me, but she never showed. After an hour, a security guard told me I had to leave.

I scrapped around for freelance writing jobs. I reviewed Greek restaurants in Queens and wrote a story for *Time Out* about things you can buy for a dollar in New York. I worked part-time out of some rich lady's Upper East Side basement fact-checking a guidebook to interior decorators that she published for her friends. I applied to the Columbia University Graduate School of Journalism even though it would've taken me the rest of my adult life to pay for a single semester. I was wait-listed. I became an early adopter of Internet trolling, blasting the whores in corporate media in manifestos that my new boyfriend (the homeless Italian with the camcorder) published on his blog. If ya can't join 'em, troll 'em. I considered starting a dog-walking business. You know it's bad when scooping poop off sidewalks sounds promising.

Finally, in the dead of February when I was almost broke, I went to the Strand bookstore and bought a lightly used copy of *LSAT for Dummies*. That same night, over a dinner of yogurt and instant ramen, I got a call from John Bussey, the foreign editor at the *Wall Street Journal*.

$$\overline{6}$$

The Foreign Desk

What is this gypsy passion for separation,
this readiness to rush off—when we've just met?
—MARINA TSVETAEVA

NEW YORK CITY → TOKYO → DES MOINES, 2004–2007
The legend went that on September 11, 2001, after the second plane hit the World Trade Center, across the street the *Journal*'s newsroom filled with smoke and debris. Security forced everyone to evacuate, concerned that if the Twin Towers collapsed westerly, they would take the *Journal* with them. John Bussey refused. As reporters and editors fled to safety, Bussey crawled under a desk to report, waded into the street to talk to people, and wrote a story.

I reread his lead every September 11. "If there's only one sight I'll remember from the destruction of the World Trade Center, it is the flight of desperation—a headlong leap from the topmost floors by those who chose a different death than choking smoke and flame."

Four months later, Bussey flew to Karachi where he met Danny Pearl's pregnant wife, Mariane, and tried to negotiate the release of Pearl, the *Journal*'s thirty-eight-year-old South Asia bureau chief who'd been kidnapped days earlier. Instead, Bussey ended up having to watch the video of Danny's captors beheading him.

Bussey was based in Hong Kong when he called to interview me to be the *Journal*'s foreign news assistant, based in New York. I don't remember exactly what he said on that first call, but he was manic and fast-talking and unimpressed with my unimpressive résumé. My name had only gotten to Bussey because of Blythe. I was a rover filling in as the assistant to Tom Wallace, the editor in chief at *Condé Nast Traveler*, until he found a permanent assistant. I'd, of course, applied for the job, but a senior editor who looked at my credentials shook her head and said, "Yeah, Tom really wants someone who went to Harvard."

I took one look at Blythe's résumé—Exeter, Harvard, classical pianist—and decided I hated her. She got the job. Six months later, we skipped out on some PR luncheon our bosses made us cover to go shopping at Loehmann's. Blythe taught me how to pull an Hermès scarf out of a tangle of marked-down DKNY and has been a dear friend ever since. A few months later, she left *Traveler* to be the *Journal*'s foreign news assistant and in typical Blythe fashion would, a year later, get promoted and move to Hong Kong. She put me up to replace her.

I didn't tell Bussey that I needed this job to save me from a pooper-scooper or worse, law school. I told him I spoke fluent Spanish and had lived in Mexico and had spent a year studying Chilean poetry in Santiago. I talked about how I'd written for the *Daily Texan* and gotten a grant to report from Chiapas on the Zapatista uprising. We got into a heated debate about NAFTA and Pinochet. He'd read my clips, even the blurbs I'd written for *Time Out*. This led to an in-depth discussion about where to find the best *saganaki* in Astoria. I'd left out the part about a prominent literary agent firing me after four months. I remember Bussey being intrigued by my impoverished New York conditions. He asked if I had a TV or AC unit. (I had neither.)

Three weeks and a couple interviews later, Bussey's deputy in New York, Lora Western, called to offer me the position.

The second I stepped into the *Journal* newsroom, nearly three years after I'd moved to New York, I felt I'd finally found my people. It was grimy, with low-hung ceilings and reams of printer paper piled up indiscriminately around an open floor plan and pea-green carpet. I arrived at the office every morning before eight. The women wore orthopedic sneakers with skirt suits, and the men all had sweat circles under their pocketed dress shirts. My banana clip fit in just fine there.

Bussey being based in Hong Kong meant I had to be prepared for middle-of-the-night requests to overnight a stash of surgical masks when the stores ran out during the SARS epidemic or to get bulletproof vests to reporters in Beirut or to find Bussey (a notorious cheapskate) a room in Davos for under a hundred dollars that wasn't a youth hostel.

I once had to take dictation from the Latin America correspondent José de Córdoba as he hid in a bathroom in Haiti during the 2004 coup that ousted Jean-Bertrand Aristide. I had to get our Middle East correspondents enough cash to buy an armored BMW in Baghdad. To do this, the *Journal* wired $40,000 into my personal checking account, which had about $400 in it at the time. I went to Chase to withdraw the cash. I still don't know if the bank representative was kidding when he asked if I wanted the money in a briefcase handcuffed to my wrist. I then went across the street to the Western Union and after some odd looks and a lot of paperwork that somehow got me around federal rules for international money transfers, managed to wire the cash to Iraq. I'm certain this errand landed me on a terrorist watch list.

But mostly Bussey and Marcus Brauchli, then the national news editor, and the other senior *Journal* editors encouraged me to blow off my assistant duties. "You want to be a writer, you have to write," Bussey said.

Marcus once saw me struggling over a pile of newsroom expenses to process and staged an intervention. He picked up a handful of receipts from various countries. The currency conversions alone were making my head spin. By then I'd written several features, including one about people who impersonate Navy SEALs, and helped with 2004 campaign coverage. But I hadn't yet cracked the front page. "You are not allowed to file a single expense report until I see your byline on page one," Marcus said. "That's an order."

He approved a six-night reporting trip to Iceland so I could write a feature story that ran on the cover of the Media & Marketing section. I wrote half a dozen A-heds—the *Journal*'s term for the quirky middle column that runs daily on the front page. One was about women under-going surgery to reconstruct their hymens so they appear to be virgins again, a story I'd come across while browsing the ads in the back of *Hoy*, the Spanish-language newspaper. I took the F train to Queens to meet Esmeralda Vanegas at Hymen RidgeWood Surgery. "It's the ulti-mate gift for the man who has everything," one of her patients told me. (An editor proposed the headline ONE NIGHT STRAND; we went with VIRGIN TERRITORY.) And I wrote a couple leders—*Journal* parlance for the serious economic or business story that runs in two columns on the front page.

I also left a delegation of Brazilian diplomats waiting in the lobby so I could finish an interview. I was on deadline and forgot to order coffee and Danish for a meeting with top editors and a dozen Alibaba exec-utives visiting from Hangzhou. I had about 200,000 rupees worth of expenses piled up that I needed to process. I'd screwed up the conversion and thought this was about US$150, but turned out to be more like US$3,000.

I killed it at being a bad assistant.

Bureau chiefs who worked for Bussey used to warn me that I should always check for my wallet when I left his office. I didn't know what they meant until he offered me a job as a foreign correspondent based in Tokyo. After our talk, I felt like a rock star, the euphoria lasting about four minutes until the practicalities of this position set in. I didn't speak Japanese. The little I knew about Japan I'd picked up from a couple Murakami novels and a five-day vacation in Tokyo to visit my friend Aika, a man-eating deejay whom I'd backpacked around South America with in college. I could hardly afford to live in New York on my salary, let alone Tokyo.

But I was cheap. The *Journal* gave me a salary raise to $40,000 a year, and called it "a suitcase relocation." I had no family or house full of belongings to relocate. I could pick up in the same red suitcase I'd used when I first moved to New York.

My mom cried when I told her. "No, no, I'm really happy for you," she said through sobbing so loud I had to hold the phone away from my ear. "But JAPAN?!"

A couple of weeks before my big move, Ellen Byron, who covered retail at the *Journal*, urged me to go out with her on St. Patrick's Day.

"C'mon, you'll meet an Irish guy," she said.

She knew that my fling with an Iraq correspondent who called me *habibti* and had a fiancée in Washington had imploded. I don't know if I even liked him that much, but I'd been so impressed with his career and his dispatches from war zones that I became infatuated. I confided this to one of my bosses, a page-one editor who shook her head of short gray curls. "Amy," she said. "You can't fuck the copy."

I agreed to go for one drink, as long as I didn't have to wear green or talk to anyone wearing a KISS ME, I'M IRISH T-shirt. There in the Pig 'N' Whistle, the type of generic Times Square pub I would never normally go to, I saw Bobby. He was brooding against the bar, a trio of blondes forming a half circle around him. He wore a khaki trench coat and had the gentlest hazel eyes framed by sad, expressive lines. With a head of thick black hair, he stood a good foot taller than the blondes. He drank a freshly poured pint of Guinness, licking his upper lip to remove the frothy white mustache that formed after each sip.

I tugged at Ellen's arm, subtly pointed his way, and said, "See that guy right there? He's my new Irish boyfriend." He was a friend of a friend, Ellen said and introduced us. Bobby told me he was from a town called Trim, "about forty-five minutes outside Dublin."

"Isn't the whole country forty-five minutes outside Dublin?" God, I was awful at this.

·But he seemed amused. We all left the bar to see a U2 cover band called Unforgettable Fire. Bobby bet me five dollars that the final song would be "Where the Streets Have No Name." I agreed, not knowing that he'd already looked up the set list.

When the concert ended and everyone spilled out onto the sidewalk, I handed Bobby a five-dollar bill. He wrapped my navy peacoat around my shoulders and flagged me a taxi. He was drunk and his upper lip quivered. We hugged goodbye. I'd just met him, yet I had the strangest feeling of my feet elevating off the sidewalk. Or maybe it was the sidewalk that had sunk beneath us. Either way, I looked up at him, disoriented, feeling punched in the gut, that place where the heart is.

"CHOZICK, BUSSEY," THE voice said before I could say hello. He always ran the words together into one Slavic-Norman surname. *Chozickbussey.* "How'd you like to go to Iowa to cover Hillary Clinton?"

It was 2007 and Rupert Murdoch had just paid $5 billion for the *Journal*'s parent company, Dow Jones. Management, unsure what the takeover would mean for the paper's political coverage, needed to install a loyal *Journal* lifer as Washington bureau chief. And there was no one more loyal than Bussey. Within days he'd left the *Journal*'s Hong Kong bureau overlooking Victoria Harbour, to move to a dreary office building off K Street. There were no niceties with Bussey. No explanation of what the job would entail or what I'd be paid. (The same salary I'd made when I left Japan, $43,000 a year.)

I'd only moved back to New York from Tokyo a couple of weeks earlier. A squishy brown Kapibara-san charm, a plush version of the world's largest rodent, still hung from my cell phone—one of my many accessories that seemed appropriate for a twenty-eight-year-old foreign correspondent in Tokyo but that I should've retired to a landfill when I got back to the States.

Bobby and I had done long distance for nearly two years. We'd lose entire days (or nights, depending on who was in which time zone) talking via a scratchy Skype connection. I'd stay in the *Journal*'s bureau late enough to call and wake him up for work each morning, and he'd do the same when the sun set in New York. I had a Miffy calendar on my desk and would mark off the floppy-bunny-eared days until his next visit to Japan or my next trip to the States.

I hadn't intended to do long distance with the Irish guy I'd picked up in a bar on St. Patrick's Day weeks before my scheduled "suitcase relocation" to Japan. I also had a rule about not dating guys who worked in finance, but Bobby was more math geek than Wall Street wolf. We shared the same gypsy spirit and even when we lay in bed my last night in New York and I confided how terrified I was to move to Tokyo—to leave everything familiar—Bobby only held me tighter, told me that I had to do it and that he'd come visit.

We fell in love in Japan. Without Bobby, Tokyo was soul-crushingly lonely. With him, we laughed ourselves silly walking through Yoyogi Park. We gawked at the girls in their Bo Peep dresses and the grown men swing dancing, their hair in exaggerated slicked-back pompadours. We practically bankrupted our favorite all-you-can-eat *shabu-shabu* joint in Shibuya and soaked in Japanese hot springs so long our skin shriveled, Bobby's cotton *yukata* hitting knee length on his sinewy six-foot-one frame. "I think we should shack up when you get back to New York," he said one night.

On his last visit to Japan, I watched him descend the elevator at Narita airport, in the ratty sweater that he'd had since his University College Dublin days. I pressed my hand against the glass partition trying to touch him and watching him disappear into the foreigners' immigration line. I turned to walk back to the Narita Express, as I always did, but instead collapsed crying outside the duty-free mall. This led to two police officers asking if I was okay. ("*Genki desu ka?*") The Japanese don't do public displays of emotion, especially not that close to the Louis Vuitton. I had to get back to New York.

Six weeks later, Bobby moved his meager belongings into the yellow-walled, rent-stabilized East Village apartment that I'd sublet while I was in Japan. The apartment overlooked a courtyard and a Hare Krishna temple with its cloud of patchouli that made our hallway smell like a hippie's armpit. Randy Jones, the cowboy from the Village People, lived upstairs and transformed the hot tar of the rooftop into an illegal garden with vines of plump tomatoes climbing up a cinder-block canopy. We were finally nesting.

Bobby's office wasn't far from the *Journal*'s newsroom. He stood next to me on the sidewalk in lower Manhattan when Bussey's call came in. It was October but warm. We had planned to walk home after work together, maybe getting dinner in Little Italy on the way, the kind of low-key evening activity that normal couples do all the time but that we'd fantasized about in so many long-distance calls. He saw my face and mouthed "What?" rolling his eyes in anticipation.

I didn't have time to explain. Bussey's tone sounded panicked, like the time he'd tried to stop Blythe and me from going to Vietnam in the middle of the bird flu epidemic. He'd framed the idea—"How'd you like to go to Iowa to cover Hillary Clinton?"—as a question, but there was only one answer.

"When do I leave?"

7

"Scoops of Ideas"

A zealot cannot be a good cultural anthropologist.
—RUTH BENEDICT, *THE CHRYSANTHEMUM AND THE SWORD*

DES MOINES, 2007

Unlike most campaign reporters who descended on Des Moines each presidential cycle and for all the steak fries and state fairs in between, I'd spent the prior couple of years covering Japan's consumer culture. I wrote a front-page story about how Westernized diets were causing young Japanese women to have larger breasts (headline: DEVELOPING NATION). In 2007, as my competitors were meeting campaign sources at Centro (CHEN-trow), Des Moines's hottest restaurant (though there wasn't a lot of competition), I was clubbing in Shinjuku with Ken-san, a Japanese deejay friend who went by the stage name Intelligent Milli Vanilli, a phonetic challenge for the Japanese. I didn't know who ran John Kerry's 2004 campaign. I'd never heard of *Politico* or its *Playbook*. The name Barack Obama sounded only vaguely familiar. When Bussey asked me to go to Iowa, I thought *for sure* I would be riding the Hillary Clinton beat all the way to the White House.

Years later I confessed to one of The Guys that when I got to Iowa, I didn't know what a caucus was. "Don't worry about it," he said. "We didn't know what a caucus was either."

I thought it would be a relief to report in English again, but I still didn't entirely speak the same language as the people I was covering, especially

Mark Penn. Penn was Hillary's trusted pollster who, after her third-place finish in the Iowa caucuses, went from being the brains behind the former first lady's political ascent to the asshole responsible for everything bad about the 2008 campaign. (Poor bastard couldn't even blame the Russians.)

Mark had some terrible ideas (like his early reminder before the 2008 campaign that "being human is overrated . . ."), but he always saw Bill's base of white working-class men as a central part of Hillary's victory. And he turned out to be correct that most voters didn't want to elect the "First Mama."

On the flight from Des Moines to Manchester after Hillary came in third place behind Obama and John Edwards, Penn pushed his combed-over sweep of coarse brown hair over his sweaty forehead and told us, "We're in a strong position to move forward." I whispered to Anne Kornblut, of the *Washington Post*, who sat in the bucket seat next to me and pounded so hard on her laptop that her tray table vibrated, "Is that really how they talk?" Anne smiled at me, as if I were a yapping lapdog that she wanted to silence. She went back to transcribing Penn's comments.

One time in Japan, during an interview with a high-level executive, I asked my interpreter, Ayako-san, to grill him on a question I knew he was evading. She cautioned, "Amy-san, it is extremely rude to ask the same question twice." I was always inadvertently being rude in Japan, so I told her to go ahead. The executive's eyes bulged. He leaned in close, his breath smelling of cigarettes and ponzu sauce. "Amy-san, I have already answered that question," he said in clear but heavily accented English. "My answer is neither yes nor no."

I remembered that exchange when I first started covering American politics, and whenever we pressed Hillary and her top aides, I imagined them all morphing into anime versions of themselves, with tiny bodies and round bulbous heads, a floating thought bubble on top: *We're in a strong position to move forward. My answer is neither yes nor no.*

EVERYTHING SEEMED LIKE a story to me in 2008. I wrote a feature about campaign hookups, a topic so baked into the process that no one thought Secret Service guys (motto: "Wheels Up, Rings Off") ducking into reporter's hotel rooms was news. I even convinced one of The Guys to talk on the record about hookups.

When everyone gushed about how nice Iowans were, I caught up with a Des Moines alt-rock band that had gained local fame for venting about the out-of-state media and political elites in a musical number called "Get Outta Our Town (Caucus Lament)." The chorus went, "Get outta our town / Get out of my face / You barged into our home / With your political race."

By the time Obama was sworn in in 2009, the Murdoch regime had started to exert its influence at the *Wall Street Journal*. The new owner-ship, judging our feature stories as having the "gestation of a llama"—or about 350 days—infused the newsroom with a new metabolism for breaking news. But during the campaign my editors still urged me to look for what they called "scoops of ideas," the offbeat feature stories that nobody else covered. I didn't realize how pretentious this sounded until I tried to explain the idea to a *Politico* reporter who could barely look away from his BlackBerry for the twenty seconds it took to reach into the innards of the campaign bus and pull out our luggage for the night.

"Did you even file anything today?" he asked me as he fished out his roller bag.

"No. My editors don't really want daily stories," I explained. My inner voice nudged me, *Don't say "scoops of ideas,"* but I didn't listen. "They want 'scoops of ideas,' you know, like rather than writing 'this happened today' or getting something from the campaign that's inevitably going to get out anyway, finding a totally different angle that no one else thought of . . ."

"Uh-huh, you go ahead with your 'scoops of ideas,' and I'll be over here breaking news," he said.

What I didn't tell him was that "scoops of ideas" were my only op-tion. I hadn't yet developed the killer instinct to compete for news, and my editors didn't seem to care when the *Times'* Pat Healy or *Politico's* Ben Smith scooped me. Not that I could've competed. I hardly had any sources. Unlike my competitors, most of whom had come up covering New York politics or Congress, no one had heard of my byline or me.

Nor did Bussey's installing me on the Hillary beat sit well with some of the paper's more seasoned political reporters. By the time I'd switched to the Obama bus in the spring, my counterpart on the McCain campaign staged an intervention. She pulled me aside outside the *Journal*'s workspace at the Democratic National Convention in Denver. We'd previously teamed up on covering a debate during which she reamed me out for not grasping the nuances between Obama's and Hillary's healthcare proposals. ("Her plan *is* a mandate. Repeat slowly after me, *MANdate.*") We'd been paired up again, this time to write the front-page story about Hillary's speech that night in which she would ask her delegates to unite behind Obama. Like Hillary, my coauthor felt the need to extend a strained show of unity to a less experienced colleague.

"Look, I wanted to say I'm sorry. I know I've been really mean to you," she said, blinking rapidly behind the foggy lenses of her glasses as we stood in the August heat against a chain-link fence adjacent to the Pepsi Center's parking lot. "It's just that I really don't think you're qualified to be doing this job."

The only person on the Hillary campaign I really got to know was Jamie, the dutiful press wrangler who remained catatonically upbeat even after four hours on a bus with reporters asking her nonstop questions about when we'd get lunch and whether we'd get our Marriott points and "Jamie, can you pass the granola bars back here? Not those, the peanut butter ones . . ." and "Jamie, I think I left my power cord in Sioux City . . ." and "Jamie, how long 'til we get there?" She was so smiley and obedient that behind her back The Guys called her the Golden Retriever.

Jamie led us into my first Hillary town hall, held in the Shenandoah fire station. Tom Petty's "American Girl" blasted from the speakers. I soaked up the Americana. There were homemade pies set up on folding tables against corrugated metal walls and red fire trucks and old men in patch-covered garrison caps. Their wives wore Christmas sweaters with puffy-paint Rudolphs and tiny flickering Christmas lights. Hillary appeared at the front of the room, and before she even said anything, I stood up from my seat and clapped. That's when I felt Jason George of the *Chicago Tribune* tugging at the right side of my parka.

"Dude! Dude!" I'd just met Jason that day, but he had the concerned expression of an old friend saving me from swallowing a handful of sedatives. "What the hell are you doing? You can't do that."

I looked at the rest of the press, all staring stone-faced at their laptops, too focused on their screens to notice my faux pas. I quickly sat back down in my chair.

8

"Taking Back America"

Writing for the *Wall Street Journal* with the nation on the cusp of the 2008 financial crisis came with some built-in advantages. Although I still dressed like a Japanese teenager—meaning I wore everything in my closet all at once, jeans under dresses, under blazers, over cardigans until I was one chubby gaijin layer—Hillary, and those on her campaign staff, assumed I was policy minded and serious. Even other reporters would frequently turn to me to ask what Hillary meant after she'd shout at rallies "*Mortgage-backed securities!*" and "*Sub-prime lending!*" All of this despite the fact that I couldn't tell you what equity derivatives were. "No one knows what derivatives are, that's the whole point," the *Journal's* finance editor once assured me.

The candidates actually wanted to do interviews with me. I got a forty-five-minute sit-down with Hillary, my longest ever. She predicted the housing crisis, warned that the US could slip into a Japanese-style "malaise" (something I did know about), and criticized NAFTA, the trade deal her husband signed into law in 1993 that would dog her through both of her presidential campaigns.

"There have been some very positive results of trade [but] . . . there is still too much of the benefits of trade and the global capital markets favoring elites and multinational companies in a way that is not spreading prosperity," she told me and my *Journal* colleague, the economics writer, Bob Davis. Bob and Hillary knew what derivatives were.

A couple of weeks before the caucuses, the *Journal's* politics editor Jake Schlesinger called. "Edwards wants to sit down for an in-depth interview about the economy after his rally today. Can you get to Vinton by two p.m.?" Jake asked.

I hesitated. I'd been in Iowa long enough to know Vinton was in the Cedar Rapids metropolitan area (if you could call 255,000 people a metro area), a two-hour drive from Des Moines on a good day. But on that day, a blizzard had parked over the state making driving conditions perilous for locals and a particular death trap for a transplant New Yorker driving a rented Hyundai Elantra with no snow tires.

"This would be exclusive. He asked specifically for the *Journal*," Jake said, in a tone neither pushy nor impolite but that told me I didn't have a choice.

"Leaving now," I said, and started to pull on my army-green parka and snow boots.

The Edwards campaign was in crisis mode after the *National Enquirer*'s JOHN EDWARDS LOVE CHILD SCANDAL! story broke. We'd all been too polite to follow the story, maybe because Edwards's wife Elizabeth had cancer or because we all thought ourselves above chasing the *Enquirer* or both. Still, for Hillary there exists an alternative route in the Rube Goldberg of why she will never become the FWP: The media runs with the Edwards baby-daddy scandal, causing him to drop out before the caucuses, allowing Hillary to pick up enough of his supporters to win Iowa, halting Obama's momentum before it started, and allowing her to win the nomination and defeat John McCain.

Trucks skidded off the interstate. I could hardly see the road and felt my tires floating on a layer of ice and snow. By the time I arrived in the cafeteria of Vinton High, Edwards was finishing his tirade against the growing divide between the haves and the have-nots. I wouldn't appreciate Edwards's "Two Americas" and his populist pitch decrying inequality and global trade until 2016 when Bernie and Trump used an almost identical playbook. Edwards ended up an imperfect messenger, a slick millionaire trial lawyer with a love child. *We know, we know, you're the son of a millworker.* But he was ahead of his time.

I found an empty seat next to the *New York Times*' Julie Bosman. Julie was the paper's Edwards beat reporter until, as a *Times* editor liked to say,

"Her horse didn't just die. He got caught fucking Secretariat." Edwards concluded his Vinton speech by pointing to our thinly populated press area.

"You see all those reporters in the back?" he said, in a gesture that felt like a game-show host breaking the fourth wall. A scattering of heads turned around. I met eyes with an older white man in denim overalls and a purple-and-gold SEIU button. "They'll be writing 'He said, she said,' while we're TAKING BACK AMERICA!"

He was even ahead of his time in shaming the elite media.

I approached Edwards backstage as his press secretary had instructed and extended my hand to shake his.

"Hi, Senator, I'm . . ."

Edwards glanced briefly at me and kept walking. "Just a second honey," he said, flashing a palm at me in a halting motion. "I got an interview with the *Wall Street Journal*."

"I am the *Wall Street Journal*," I said.

9

Leave Hillary Alone

I knew Hillary was running again on a Monday afternoon in early May. She'd sent all The Guys, including Outsider Guy who avoided DC and had to fly cross-country, to attend the meeting, along with her closest aides. These included Cheryl Mills, the classy, elusive lawyer who'd defended the Clintons during impeachment and Benghazi and who knew how to drape a scarf in even the hottest State Department convoys to Senegal; Tina Flournoy, a former union leader and boxing aficionado from Georgia whose combination of tenacity and Southern charm made her uniquely qualified to hold the unenviable role of Bill Clinton's chief of staff; and Huma Abedin, the elegant waif and tabloid fixture who'd worked for Hillary since she was a nineteen-year-old White House intern and George Washington University student.

Ever since I arrived in Iowa in 2007, I'd marveled at Huma the way women tend to marvel at impossibly thin, fashionable women. She was the only one (including Hillary) who didn't gain at least ten pounds in 2008. It got so bad that by the Indiana primary, I saw Chelsea swat her mother's hand away (*"Mom!"*) from a deep dish of chips and salsa. And I watched, hardly able to keep up in my bulky snow boots, as Huma glided alongside Hillary in stilettos during the Scranton St. Patrick's Day parade. The press speculated about whether Huma had hooked up with one of The Guys who had a fiancée in New York, our very own soap opera unfolding on the plane. But on one flight toward the end of the primary, Huma introduced the '08 traveling press to her new boyfriend—a promising young congressman from New York named Anthony Weiner.

It had been just over a year since Hillary had stepped down as secretary of state, and now certain that Carolyn Ryan had inherited the personal vendetta against her family, she had instructed her most trusted loyalists to convene at the *New York Times*' Washington bureau and express her concerns about my coverage. Why would Hillary do that if she wasn't running?

When The Guys emailed us a list of the seven aides who planned to attend the meeting, Carolyn wrote back, "We're gonna need a bigger boat."

We all agreed the discussion would be off the record, but it didn't take long for somebody to tip off the conservative *Washington Free Beacon*, which soon after published a story with the headline HILLARY TO NEW YORK TIMES: BACK OFF. But the *Beacon* story only mentioned the presence of Original Guy and Huma—who was hard to miss floating through our slovenly newsroom like an exotic bird in a red wool coat. They didn't know the half of it.

Everyone huddled into the narrow entryway under the bureau's fluorescent lights. The Guys forced themselves to offer me their usual clipped hello, which always reminded me of Seinfeld opening the door for Newman. "Hello, *Amy*."

After some handshakes, Carolyn led everyone down the drab, carpeted stairway to the conference room where The Guys helped themselves to coffee only to find that it had been left over from a previous meeting. The brown liquid was dank and acidic, and the creamer crumpled as it splashed into their Styrofoam cups. We hadn't planned it that way, but serving stale coffee and day-old Danish to DC's most powerful people did send an effective message that things probably wouldn't go their way.

I can't get into the details. I can only say that they griped about stories I thought were positive, like one about Bill building a charitable legacy in Africa. (They hated the timing.) They complained about stories I thought were neutral, like Hillary working to rebuild bonds with black voters. (Black people never left the Clintons, they said.) They understandably despised a story about a Ukrainian oligarch, Victor Pinchuk, who'd given millions of dollars to the Clinton Foundation and had a slew of meetings with State Department officials. (Victor liked the story. He invited me to his annual conference typically held in Yalta, the tsars' fabled Black Sea resort town. I declined.) But mostly, they hated that the beat existed at all. They said Hillary was a private citizen.

The story that put me on the radar as the *Times'* Hillary chronicler and made that proverbial target on my back more like a permanent tattoo arrived in a January 2014 issue of the *New York Times Magazine,* along with a doughy Hillary moon face floating amid intergalactic dysfunction on its cover. I'd meant for my "Planet Hillary" story to serve as a fifty-seven-hundred-word primer about all the people the Clintons had collected over the years and the "organizational meshugas [that] already threatened, once again, to entangle" Hillary as she prepared for 2016. In an accompanying chart, I categorized the Clintons' minions (and almost all my would-be sources) into competing solar systems. There was "The Inner Circle," "The 2008 Victims," "The People Who Do All the Work," "Loyal Henchmen," "Frenemies," "Poseurs," and so forth. Not surprisingly, almost everyone hated their designated place in the universe.

Doug Band called to complain that he'd been positioned (in the category "The White Boys") next to the floating head of the irascible wonk and former White House policy adviser Ira Magaziner. "I can't believe you put me next to that asshole. You know I hate that guy!"

One of the Poseurs yelled at my editor that the placement could've cost him his gig as a paid Fox News contributor. Another offered to get Bill Clinton on the phone to tell us that he wasn't a poseur. To which my editor replied, "That's exactly what a poseur would do!" A New York executive told me the "Poseur" label had been the "worst thing to ever happen to me." (He lived a charmed life.) Worse, I'd been somewhat of a puppet in all of this, later learning that all the poor schmucks who ended up designated as Poseurs had at some point pissed off The Guys, who'd accordingly steered me toward tagging them with that label. One of them, a friend and donor, was also the ex-husband of the buxom blonde whom one of The Guys had an affair with on an earlier Foundation trip to Africa.

I'd started to get used to the idea of breaking some eggs to make an omelet, but with the "Planet Hillary" story, I'd dropped the whole damn carton.

Carolyn always had her reporters' backs. I knew that after the DC meeting, she'd plop down on the sofa in her office next to the Ping-Pong table and tell me to never doubt my coverage. She'd remind me that we should have a combative relationship with the people we cover—and she'd say that I deserved combat pay. But during the meeting, Carolyn didn't say much. She disarmed the group the same way she disarmed reporters who came into her office unprepared. She took exaggerated sips of Diet Coke and squinted as they spoke, sometimes jutting her neck toward our visitors and then leftward to me giving the impression that she was listening to crazy talk and craved simultaneous translation.

I kept trying to fill the silences. I apologized. I said I'd try to do a better job next time and I'd be more careful moving forward. But that just pissed The Guys off more. The shrinking violet act and all.

They all seemed trapped in a time warp. Whitewater was yesterday, but all the positive stories and endorsements the *Times* had given Hillary in recent years were worthless relics.

I could've tried to defend myself, but I was up against seven much smarter professional Hillary defenders, including Cheryl, a former deputy White House counsel. I didn't stand a chance. I kept thinking of the scene from *Full Metal Jacket* when Matthew Modine's Private Joker says, "Sir, the private believes that any answer he gives will be wrong and the senior drill instructor will only beat him harder if he reverses himself, Sir!"

My train back to New York after the meeting was delayed for hours. I crouched on the floor of Union Station between a Jamba Juice and an Auntie Anne's pretzel to charge my iPhone and check my messages. The Guys had dumped me. Their email might as well have said, "It's not you, it's me." Or they could've used the same line the Colombian waiter I went on a couple of dates with in my twenties did when he texted me, "U R 2 high maintenance."

From then on, I was to deal directly with Cheryl.

Though she outranked The Guys, this didn't feel like a reward. My job required me to have some semblance of a relationship with Hillary's press aides. There wasn't even a campaign yet, and I'd already failed. As it turned out, my interlude with Cheryl lasted a couple of weeks until she got sick of me, too, and kicked me back to The Guys. Only this time instead of the OG, I was now to deal with his less experienced but more presentable protégé, the brown-loafers wearer.

Months later, when OG presented his Mini-Me to the rest of the political universe, everyone thought of him as the nice one. They'd gush, "Have you met him yet? He is *such* a nice guy." And at first, Brown Loafers Guy was a breath of fresh-faced air. He burst with optimism about Hillary's future and his own. He had OG's biting wit and a full suite of adorable facial expressions that played well on cable TV. He had a direct line to Hillary (who adored his adorableness) and none of OG's dark edge, sexist undertones, or tendency toward high drama.

The Guys got help from outside supporters. A ragtag group called the HRC Super Volunteers sent me a warning: "We will be watching, reading, listening, and protesting coded sexism . . ." According to their list, sexist language included "polarizing," "insincere," "inevitable," and "secretive, will do anything to win, represents the past, out of touch . . ."

David Carr always had my back. Like the blunt conscience of the *Times*, David proved the only person who could really defend me. "HRC's minions throw brush back pitch at NYT. Look for NYT to lean in and hit one hard up the middle," he tweeted after news of the DC confrontation leaked.

He'd tell me again and again, it's not you, it's them. "You never made a single enemy on the media beat," he'd say.

Sometimes, when I needed an extra confidence boost, he'd email me one of his David emails. "There is no one else like you," he wrote. "Doubt yourself as a writer if you need to—it will drive you to new ways of thinking—but don't doubt that. You are your own damn thing." Despite writing a weekly column, mentoring Lena Dunham, and helping out on every breaking news story he could get his tarry hands on, David still made time to bestow emails like that (and corresponding spirit animals) on a small army of younger journalists.

Around the same time, The Guys took smug satisfaction in the *Times'* abrupt firing of Jill Abramson, a development that had nothing to do with Hillary coverage and that left me, like many young women in the newsroom, floored and sad. David would swing by my cubicle, a scarf wrapped in a Parisian knot around his pencil-thin neck, crumbs from his morning donut stuck in the crevices. He wouldn't say anything. He'd just make a claw motion with his hand and growl, a reminder that I was the Polar Bear.

But polar bears are also lonely and endangered. I was floating on my own little iceberg, and it was melting fast.

MY INTERACTIONS WITH Hillary over the course of 2014 continued to be few and far between, usually chance encounters when she'd always pretend to be *thrilled* to see me.

In the spring, Bobby and I went to the premiere of a documentary film that Chelsea had executive produced about the unlikely friendship between an imam and a rabbi. It wasn't exactly the red carpet event of the century, and I turned out to be one of the only reporters there.

Halfway through the cocktail party, Hillary walked in and made a beeline for the bar. Chelsea had announced earlier that day that she and her husband, Marc Mezvinsky, were expecting their first child. I no longer saw myself in Chelsea. She had grown into her celebrity, with flowing, straight hair and a permanent strawberry glow. Chelsea told *Elle* magazine that in her early twenties, her curls just naturally subsided, an affront to frizzy-haired women everywhere. I also happened to know her New York hairdresser—and a keratin job when I saw it. Chelsea's press aide told me they'd studied how Britain's royal family had handled Princess Kate's pregnancy to devise the media strategy.

"Congratulations! Such wonderful news. How excited are you to be a grandma?" I said, sidling up to Hillary at the bar. I put my hand on her shoulder and felt the luscious satin of her chartreuse tunic beneath my palm.

Hillary took a sip of pinot grigio and as she swallowed said, "Oh, Amy, it is just the absolute best."

We walked into the crowd. "Secretary, I'd like you to meet my husband, Bobby," I said.

Bobby, the oldest son of Irish school teachers, is from County Meath, a sweep of fluorescent green farmland on the River Boyne. The Trim Castle, a grand Norman structure used as the backdrop of the movie *Braveheart*, stands blocks from his family's redbrick house.

Like many Irish, he has a special place in his heart for the Clintons and their commitment to the peace process. He has hazy childhood memories of the British army shoving their guns into his parents' Datsun Bluebird when his parents would drive across the border to Belfast. I picked up early on that the best way to get on my mother-in-law's good side was to declare something Irish superior to its English equivalent. "The brown bread just tastes better in Ireland." Or, "Why can't an English breakfast come with black and white pudding?" I learned the Irish words for Christmas sweater, *geansaí Nollag*.

Bobby had hardly said hello when Hillary interrupted. "Is that an Irish accent I detect?" she said.

They tucked into a corner (out of my earshot) and talked for ten minutes about the Good Friday Agreement, their mutual concern that the crash of the Celtic Tiger could reignite the Troubles. I stood there making small talk with Marc Mezvinsky, watching Hillary and Bobby out of the corner of my eye. They ended up talking for longer than I'd talked to Hillary in months (years?). I wanted to crash, but I didn't. For all the times Hillary had inadvertently interfered in our relationship, leaving them alone to chat was the least I could do.

In the taxi back to the East Village, Bobby sank down into the seat and propped his knees against the back of the Crown Vic. He isn't a talker. I usually blab, and he listens and then inserts wisdom and witticisms. But that night in the taxi, he went on and on about meeting Hillary and their conversation with the elation of relaying the time he'd seen U2 play at Slane Castle. I listened, happy to see him so happy, grateful for the reminder of that side of Hillary.

A couple of months later, at a naturalization ceremony that included immigrants from a hundred countries all waving American flags and mouthing the words to Lee Greenwood, Bobby was sworn in as an American citizen. Right after that, he registered to vote. I would try to see the 2016 election, and Hillary, partly through Bobby's uncynical immigrant eyes. "For fuck's sake, she brought peace to Ireland. I don't care if she's funny on *SNL*," he'd say during the campaign.

BY THE FALL of 2014, I thought we'd turned a corner. Or, at least, I'd learned how to handle the beat without raiding my mom's dwindling Xanax stash. There were actual events to cover. Hillary did the Harkin Steak Fry in Iowa. She campaigned for midterm Democrats. *Hard Choices*, her empty brick of a memoir about the State Department, came out. The fledgling Hillary traveling press corps trailed her to every dreary midterm rally, every Barnes & Noble and Costco book signing.

We all went to Little Rock for the tenth anniversary of the opening of the Clinton Presidential Library, which drew the 1992 campaign alumni. ("Hey, Hillary! Begala's still got his jacket," Bill yelled, pointing to Paul Begala in a denim Clinton-Gore '92 jacket embroidered with a thrusting donkey.)

The weekend included an after-party at the mansion of the Clintons' Little Rock decorator Kaki Hockersmith (known in DC as Tacky Kaki) and featuring Kevin Spacey holding court at an outdoor bar doing his Bill Clinton impersonation. And there was a late night at the Capital Hotel bar in Little Rock, where an inebriated Terry McAuliffe put his arm around me and said, "Amy, can you believe I'm governor?!" No. Gene Sperling, Clinton's verbose economic adviser, cornered me until after 3:00 a.m. to defend the earned-income tax credit. Sid Blumenthal stewed in a corner nursing a Moscow mule.

Ready for Hillary, the group that called itself a "grassroots super PAC" (as if that weren't an oxymoron) held a donor confab at the Sheraton in Midtown. James Carville, Paul Begala, and other members of the original Clinton war room held panel discussions on topics like "It's the Economy Stupid" and "Lessons Learned from 2008."

They critiqued Hillary's '08 campaign, telling reporters that "every six weeks there seemed to be a new slogan, and there was nothing people could wrap their arms around." Harold Ickes, known in the White House as Bill Clinton's garbageman for reasons that had nothing to do with waste disposal, briefed donors from a third-floor conference room. He predicted a hard-fought 2016 general-election battle in which Hillary would confront Jeb Bush–Rob Portman, a ticket bolstered by a simple message along the lines of "It's time for a change."

But the biggest precampaign schmooze fest was the Clinton Global Initiative (CGI) in New York in September (on my birthday, to be exact), the Davos-like gathering that matches wealthy donors with worthy causes. Because this would be the last CGI before Hillary became a presidential candidate, the press shop had assigned handlers to escort reporters everywhere, lest we run into a donor who went off message. The theme that year was "Reimagining Impact," not to be confused with 2013's "Mobilizing for Impact" or 2012's "Designing for Impact." There was a lot of impact happening at CGI.

I wrote a brief blog post about the young press minder (an intern, I later learned) who had followed me into the restroom. When I asked one of The Guys for comment, he sent me a press release about American Standard's "Flush for Good" campaign to improve sanitation for three million people in the developing world. "Since you're so interested in the bathrooms and CGI," he said.

It was worse than the Yorkie. It was worse than anything else I would publish for the next three years.

I'd written the potty-minder post as a brief, breezy CGI scene-setter, not a serious commentary on relations between the Clintons and the media. But that's not how the wider world saw things. The *Washington Post* published a column THE CLINTON TEAM IS FOLLOWING REPORTERS TO THE BATHROOM: HERE'S WHY THAT MATTERS. The *Free Beacon* called for one of The Guys, ironically the most decent and professional of the cohort, to "stick his big obnoxious head in the toilet and 'Flush for Good.'" That didn't help matters. Until then, I hadn't fully grasped the impact of a *Times* story in the viral news era. Bathroomgate was discussed on the *Today* show, CNN, MSNBC, NPR, and ad nauseam on Twitter. I declined every interview request. I just wanted it to go away.

By the time Bill stepped off the stage after CGI's closing plenary session (called "Aiming for the Moon and Beyond" because he spoke via a satellite link to a couple NASA astronauts who appeared, weightless, on board the International Space Station), the only story out of CGI anyone was talking about was the bathroom incident. "Goddammit, we're trying to save the world and all these people can talk about is the goddamn bathroom," was how one person summed up Bill's backstage reaction.

Hillary's expletive-laced response was worse. She told The Guys she'd held out hope I might still treat her fairly, but she'd given up on me after the bathroom post. "To be very honest, this episode was upsetting to people, not least of which the foundation team," Brown Loafers said.

The Guys told me the post and a subsequent selfie I'd tweeted with a different press minder had "humiliated" a young intern. I felt awful about the whole thing. I hadn't identified the intern and didn't know her name. I had a handwritten apology note, but The Guys (who demanded I apologize) wouldn't tell me where to send it. I could handle another fight with The Guys, but the last thing I wanted was for some hardworking kid to be inadvertently swept up in my media shit storm.

After that, The Guys and I tried to avoid one another. They'd ask if I was working with any (preferably male) colleagues or researchers and said they would "gladly" talk to them instead. Of all The Guys, Outsider Guy, who a couple years back had fought to get me access and unleashed on Ugandan military officials who wouldn't allow me (a "cockroach reporter") into a Clinton Foundation event, had become the most venomous. Maybe because he knew me the best, ever since Iowa and the time we'd shadowed Bill and Chelsea shaking hands and stirring up mayhem in Las Vegas casinos ahead of the 2008 Nevada caucuses, Outsider Guy also knew how to wound me more permanently than the others. The things he said stuck with me as I morphed, in his eyes and occasionally my own, from ally to cockroach.

On a story about Martin Scorsese killing an HBO documentary on

Bill Clinton's life after Chelsea had allegedly requested final cut, Outsider Guy would deal only with Michael Cieply, my coauthor in Hollywood and a grizzled industry veteran. "It's hard for me to believe you deal with them for a living," Cieply said, adding that his brief conversation with Outsider Guy had been the nastiest exchange of a career that had included getting yelled at by Harvey Weinstein and several studio executives sniffing coke off conference tables.

I tried to give The Guys a taste of their own medicine.

One night, at a cocktail party in the West Village townhouse of a former White House aide, a pile of Clinton hands, old and new, talked about the recent news that Robert Gibbs would leave his role as Obama's White House press secretary to be the top corporate flack at McDonald's.

"You couldn't pay me enough," one of The Guys said.

"I'd rather work for big tobacco. Seems more honest," a White House aide turned Wall Street executive agreed.

I was in a debate with The Guys about a page-one feature set for the weekend paper. I explained that this would be a heartfelt portrait of Hillary's mother, Dorothy Rodham, and how her childhood struggles would form the emotional core of her daughter's 2016 campaign.

"Really? There's nothing else I should know?" Hired Gun Guy said. "You always find a way to include some kind of dig . . ."

"You're serious? You think I'm going to take a dig at her dead mother?"

"I don't know," he said, lifting his shoulders a couple of inches and pushing his open-palmed hands out in a cartoonish shrug.

"You know," I said, taking a sip of rosé and cutting him off, "best case scenario, this all ends with a job at McDonald's."

10

"Iowa . . . I'm Baaack"

INDIANOLA, SEPTEMBER 2014

"Secretary! Can you believe you're back in Iowa?"

"Hillary! Does this mean you're running?"

"Can you win here this time?"

Hillary stood in front of a Char-Griller, pretending to flip a steak at the Harkin Steak Fry in Iowa. It was the political event of the year for Democrats, Hillary's first trip back to the state that had wrecked her 2008 presidential campaign, and the clearest public sign yet that she would run again. (For decades, the Harkin Steak Fry was the mandatory testing ground for would-be Democratic candidates.) I watched as she held the wooden handle of a spatula at a safe distance, as if a garden snake were coiled around the end. Then she gave the photographers a smile so open mouthed and amplified that, looking back on it, I should've seen it as a cry for help. The entire Democratic Establishment should've seen it. The image screamed all at once, *How long do I have to act like I enjoy this shit?* and *Why the fuck am I back in this state?* and *Dear God, what am I doing?* But what Hillary actually said to the press that afternoon was:

"It's *gurrrrate. It's fabulous* to be back. I *love* Iowa."

I LOOKED AROUND the press scrum at the steak fry. We were all scrambling to write almost identical stories, using almost identical quotes and almost identical color ("She smiled in front of hay bales, an American flag and a John Deere tractor."). The last time I'd been in Iowa with

Hillary, I still felt like a foreign correspondent. Now, my journalism had become more like a feeding frenzy than a moveable feast.

I was no longer the kid who didn't know any better than to stand up and cheer at a town hall. I'd become omnivorous, driven beyond all rationale by byline count and Twitter mentions. I lived in fear of being scooped over even the most insignificant minutia. *Politico* may have beaten me on the Ready for Hillary fund-raiser at the Standard Hotel in New York. But damn it, I heard the *exclusive* news that tickets cost $20.16 and that the signature cocktails included an eighteen-dollar gin-and-lime concoction called the Ultimate Ceiling Breaker. We were so starved for tiny morsels of news that I groveled with one of The Guys to use the names of his cats, Uday and Qusay (named after Saddam Hussein's sons "because they were little terrorists"), in the "Planet Hillary" story. I was starting to see things—even how Hillary flipped a steak—with cynicism, and I feared the coming campaign would engulf what was left of my wide-eyed 2008 self.

AT THE STEAK fry, Iowa Hillary—the most belabored of all the versions of Hillary I'd mentally characterized—delivered an unintentionally ominous "Hello, Iowa, I'm baaaaaaaack." She cracked a joke about Bill's vegan diet. "It does really feel just like yesterday when I was last here at the Harkin Steak Fry, or as my husband now prefers to call it, the stir-fry," drawing some giggles from the leery crowd of party faithful.

She then embarked on fifteen minutes of vanilla remarks. "In Washington, there's too little cooperation and too much conflict . . ." she said, in what, even for Hillary, ranked high in the pantheon of pabulum political talk. This was when Bernie Sanders was still an obscure socialist senator from Vermont, and Hillary's aides had urged her to take out any references to raising the minimum wage, advising that "this might still be too hot and partisan and might prefer just saying 'We have a choice whether to move forward . . .'"

Bill sat behind her, his mouth newly flopped open in a way that made people assume he was older and sicker than he actually was. He wore a red-and-white gingham print button-down shirt, a recent birthday present from Hillary. "It kinda makes me feel like a tablecloth at a diner," he told us.

When Hillary finished, Tom Harkin took the podium and in nine folksy words stroked Bill's fragile ego and undermined Hillary in a scene that stayed with me for the rest of the campaign. "We saved the best for last, didn't we, folks?" Harkin said. Chants of "Bill!" echoed over the grassy field.

As both Clintons headed back to their SUV and an accompanying eleven-car motorcade, a handful of young Latino immigrants, whose numbers had swelled since 2008, shouted out to Hillary about whether she agreed with Obama's mass deportations. Would she deport their families, too? She shoved a thumbs-up their way and said, "Yaaaay!"

Salon called the event THE DUMB IOWA STEAK FRY: AN OMEN FOR THE HORRIBLY DULL POLITICAL YEAR TO COME. MSNBC's Joe Scarborough said of the footage of Hillary flipping steaks with a forced grin, "Hillary Clinton's problem for people that know her and like her—like I know her and like her—she puts on that political hat, and then she's a robot."

WHEN I GOT back to Des Moines that night, I grabbed a seat at Centro at the end of a long table next to Hillary's faith adviser, a blubbery, histrionic man. We called him Hands Across America (HAA) because he'd traveled the country with us in 2008 after a sexual harassment allegation led to his brief banishment from the Hillary campaign's Virginia headquarters. A young staffer said he planted a wet, unwanted kiss on her head. (The staffer was transferred to a different department.)

After that, HAA ran an outside group that supported Hillary's 2016 bid, using his position to regularly feel up several of the young women who worked for him in hopes of landing a job on the campaign. He would be frozen out of the official 2016 campaign team.

HAA exhibited generally creepy behavior, but seemed more pitiful and effeminate than threatening, which is why I tried to ignore his rubbing up and down my back at the steak fry as we posed for a selfie that I posted on Instagram as if we were old friends. I once had a meeting in DC with HAA and a family friend of Hillary's who had the porcelain smile and abundant black lashes of a daytime TV host. HAA rubbed his hands together and said in his Southern drawl, "Ay just luv my job. I get to be in a locked office with all y'all pretty, young girls." Hillary would sometimes mention him on the campaign trail, referring to her "friend" who "sends me scripture and devotionals, sometimes mini-sermons every day," always leaving out the small detail that everyone suspected he was a pervert.

As I took my seat next to HAA I tried to ignore this detail, too, because that's what political reporters do when we are in Iowa. We write identical stories and suck up to drunk, lecherous sources at Centro.

HAA massaged my shoulder with one hand and drank a whiskey on the rocks with the other. I put a couple of the fried brussels sprouts with ranch dressing on my plate and sat there without speaking, attempting to contort my face into the expression I thought an unbothered male reporter would make.

The other reporters grilled HAA about when Hillary might declare and whether she would even have a primary opponent. He pretended, like everyone on the unofficial payroll then, that she hadn't made up her mind.

"Now, c'mon y'all, give her the space to make up her mind. She wants to take her time, do it right this time. I sent her a scripture this morning that said . . ."

I felt his hand move down my back.

"I'm not feeling great, I think I'm going to head back to the Marriott," I said, jumping up.

"No, Ames, now c'mon, you just got here," HAA said, tugging on the arm of my blazer.

I dug around in my backpack and pulled out all the cash I had, eleven dollars, and tossed it in the middle of the table.

11

The Last Good Day

HONOLULU, DECEMBER 2014
Carolyn couldn't get me combat pay, but she must've known I was about to crack because she agreed to put me on the cushiest assignment in journalism as a reward after a bruising year in Hillaryville: babysitting the Obamas on their annual Hawaiian vacation.

By my third day at the Moana Surfrider hotel in Honolulu, I had the timing down. I'd wake up at 4:00 a.m. in my corner room of the old side of the hotel, overlooking Waikiki Beach and Diamond Head, and check in with my editors in Washington. I'd file the first draft of the "setup story" with any anticipated news (e.g., POTUS's statement about the North Korean hack on Sony; his planned visit to a mess hall at the Marine Corps Base) around 7:00 a.m. This is all Hawaii-Aleutian time. Then I'd head downstairs to find a spot on Waikiki Beach close enough to where the waves broke so that the sound of the saltwater drowned out passing tourists, but far enough from the shoreline that the late-afternoon tide wouldn't sweep up my laptop and reporter's notebooks and Richard Ben Cramer's *What It Takes*, the 1,072-page tome on the 1988 election that I was determined to get through before the 2016 campaign started, even if it meant skipping the Gephardt chapters.

There were only a couple weeks left of 2014. Almost a year had passed since "Planet Hillary" made me persona non grata, and even some of the Poseurs were speaking to me again. Six months had gone by since the DC confrontation and nothing made Carolyn happier than strutting into the daily morning meeting with the *Times'* top editors and fighting to get my stories on the front page.

I'd learned not to rely on The Guys. I'd cultivated a variety of sources that ranged from one of Bill Clinton's kindergarten friends to a State Department official turned Wall Street executive. Sourcing up usually involved a friendly off-the-record breakfast (I stopped doing dinners after the "inside you" incident). A morally ambiguous donor and former aide always insisted we meet at the King Cole Bar at the St. Regis in Midtown Manhattan under the gilded Parrish mural of merry Old King Cole surrounded by his obsequious court of knights, musicians, and servants. I always wanted a shower after.

The *Times*' new executive editor, Dean Baquet, gave me his full support as Jill had. Dean is from New Orleans and wore black suits with red pocket squares and rimless glasses. He called me "kiddo" and loved to gossip about Little Rock and Juanita's, the Tex-Mex restaurant where he met sources when he was a *Los Angeles Times* reporter looking into Hillary's commodity trades. Because of this early stint in Arkansas, Dean, along with Carolyn, succeeded Jill and Howell Raines as the latest embodiment of the Clintons' theory that the *Times* had it out for them.

The bad blood from Bathroomgate never went away. I remembered something a *Journal* editor told me after a lengthy correction was appended to one of my early stories: "We are all forged in the crucible of our mistakes, both professionally and personally." That was true of me and the CGI bathroom story. I assumed it was true of Hillary, too. Forged in the crucible of all the conflicts she'd endured and the mistakes she'd made, ready to confront another campaign as an older, wiser, better version of herself.

Ever since Jill put me on the beat, I'd anticipated 2015—the year Hillary would be a formal candidate and I'd have an actual campaign to cover. But I also dreaded what was to come: the stress, the constant travel, the battles both with The Guys and inside the Steel Cage Match. Until then, the Hillary story had been mostly mine, but soon it would be bigger than me, bigger than any of us. I couldn't postpone the inevitable, but by some stroke of luck and a lot of sucking up to Carolyn, I could ring in 2015 on Oahu, one of the last places on earth to celebrate the New Year.

By 3:00 p.m. Hawaii-Aleutian time, Washington had completely forgotten about me. They'd practically forgotten about POTUS. I'd send notes updating my editors on his movements, as instructed—"Presidential motorcade departed Kailua compound at 4 p.m.," or "POTUS is bowling with friends." But Obama's mundane vacation whereabouts hardly warranted a story. So I packed every indulgent afternoon with things I knew I wouldn't have time to do in the next twenty-two months until Election Day.

I took surfing lessons. I discovered the Frosé and sucked them down like seventeen-dollar Slurpees. I tested out various shave-ice options before settling on a little stand in a back alley behind a Thai massage parlor. I never wore makeup or anything other than flip-flops and left my watch in my hotel room because I didn't want a tan line. I didn't care what time it was anyway. Bobby came to visit for a few days with his golf clubs and SPF 50.

On day seven, reinforcements arrived to help me with this arduous assignment.

The *Times*' Mike Schmidt waddled onto the warm carpeted sand of Waikiki fully dressed and squinting. With a bulky black ThinkPad under one arm and his other hand cupped over his eyes looking out toward Diamond Head, he spoke loudly into an earpiece. ("No, it has to go tonight or we lose the exclusive.") Had anyone else's restive DC ambition planted itself upon my eighty-five-degree beach day, I would've been pissed off, but the sight of Schmidty made me smile. I waved for him to come sit, but he looked right past me.

"Yo, Schmidt, over here!" I stood up.

"Hold on a sec," he said into the phone. "I didn't recognize you. You look Latin."

This didn't feel like an accomplishment. In the time that I'd worked on my tan and learned to stand up on a surfboard for a grand total of fifteen seconds, Schmidty had already written a feature about Obama's mediocre golf skills (14 handicap, at best) and had broken a real talker on the presidential motorcade tapping inexperienced volunteer drivers to shuttle the press. He'd reemerged onto the beach in a swimsuit and was talking about how he needed to "source" (i.e., drink) with White House staffers who spent most of the Hawaii trip lounging around the pool hoping not to have to partake in tropical beverages with some go-getter reporter in Panama Jacks. We were bad enough fully clothed.

I loaned Schmidt the highest SPF I had and dragged him into the ocean. We didn't so much swim as wade in to where the water was so deep we couldn't touch and buoy around—a couple of uptight beat reporters dipped into the warm water.

The surfers and catamarans on the horizon drew nearer and my de facto beach office became a tiny sliver on the sand. I don't want to call it a baptism. That would be melodramatic, and if the two of us had anything besides scoring the winter White House gig in common, it was that we looked like old friends from Jewish summer camp. But the universe was trying to tell us something in the ocean that day.

Schmidty calls it the Last Good Day. I think of it as the Afternoon of Impending Doom. Whatever you call it, we got back to the mainland and before our tans had faded, an editor called me late one Thursday night to tell me that David Carr had collapsed and died in the newsroom.

I dragged myself into the office the next morning, past the spot under the *Times* awning where David always smoked. I took the elevator to the second floor, walked to what had been David's corner cubicle—his piles of illegible notes scrawled on legal pads, the backsides of press releases, and the insides of file folders; his silly Minnesota knickknacks; donut crumbs sprinkled like a dusting of snow across his desk; that scarf strewn over the back of his chair. I sat on the floor, by the trash can, pressed my back against the cold windows overlooking the Port Authority Bus Terminal, hugged my knees into my chest, and bawled. I tried to muffle this drooling, groveling fit into my gray sweatshirt. So many people lost David—his wife and three daughters; his siblings; his journalism students; his neighbors in Montclair, New Jersey, who knew him as the goofy suburban dad with the leaf blower; the millions of readers who relied on his steady, scathing voice to make sense of things. But for those three minutes in a cubicle that still smelled of Camels and cafeteria coffee, I let myself wallow in self-pity, sobbing to the spirit of David that I couldn't make it through the election without his all-knowingness, without our ramen lunches and his reminders that I deserved to be where I was, doing what I was doing.

Two weeks after that, Schmidty broke the story that Hillary exclusively used private email at the State Department. I soon found myself at the United Nations for the first "WHAT ABOUT YOUR EMAILS?" press conference and everything took an irrevocable turn for the worse.

12

Emailghazi

You swallowed everything, like distance.
Like the sea, like time. In you everything sank!
   —PABLO NERUDA, "THE SONG OF DESPAIR"

UNITED NATIONS, MARCH 2015

The 2016 election started on the second floor of the United Nations in a tiny stretch of a midcentury hallway. It had been eight days since the initial *Times* story broke and the hysteria over Hillary's email practices had only intensified. She'd sent out an 11:35 p.m. tweet that went through half a dozen revisions and tweaks from lawyers until everyone settled on "I want the public to see my email. I asked State to release them. They said they will review them for release as soon as possible." The Guys were dizzy counting retweets.

"A thousand retweets already."

"Good to know people go to bed with their Twitter decks. Sex must be obsolete," wrote John Podesta, the veteran White House aide who would serve as Hillary's campaign chairman.

"I just spit out Diet Coke onto my desk," Brown Loafers replied. "We're at 1,791."

Six days and nearly eight thousand retweets later, we got word that Hillary would be holding a "brief press conference" after addressing a Women's Empowerment Principles event at the UN. An email from The Guys included details for "those of you that are not already coming," which included just about everyone. The alert sent several hundred reporters, cameramen, producers, and photographers to storm the otherwise sleepy UN credential office. The room had the feel of a multicultural DMV with brown twill carpet, vintage desktop computers, and the words MEDIA ACCREDITATION printed on typing paper and stuck to the wall.

A lone UN staffer seated behind a low desk tried to handle the deluge of requests to get inside—a process that typically takes weeks and requires background checks, social security numbers, passports, and recommendation letters. The Guys emailed us all at 11:27 a.m. saying we had to reply by 11:45 a.m. to secure a spot.

At the risk of revealing my complete lack of journalistic instinct, I'll admit that when Schmidt first told me about how Hillary had used clintonmail.com, I'd thought it was another one of his killer scoops, but nothing earth shattering, certainly not a "nuclear winter," as a former National Archives official called it. I think I said, "Oh wow, cool story. Thanks for the heads-up." Even after seeing the reaction to that first story, I predicted the fever would lift in a week or so, that people would get tired of reading about the intricacies of the Federal Records Act (snooze) and move on. But the story only gained speed. I'd never been a part of anything like it. The standout scandals of 2008 (Jeremiah Wright, Bristol Palin's teen pregnancy) mostly offended along ideological divides, storms that would eventually pass. But the private server was more like an avalanche, blind outrage that barreled through every newsroom and war room, devouring everything.

All this came in the midst of Bobby and my finally concluding a two-year apartment hunt, having settled on a co-op on the Lower East Side in a hulking redbrick midcentury tower that still had a Shabbos elevator that stopped on every floor.

Bobby had done all the paperwork to secure a more favorable mortgage. "Do you promise to be there? All you have to do is show up," he'd said the night before organizing our tax returns and pay stubs and contracts I'd never seen before into a manila folder.

"Yes, of course, I promise. I'll be there."

And I was there, sort of. As Fred from Citibank explained our fixed interest rates, I was also on a conference call with my editors planning how we'd cover the UN press conference. I unmuted the call.

"I can file quick off the presser if Schmidt wants to take the lead . . ." I said.

Then I muted the line and turned back to Fred. "Yep, yep, that all sounds good. Where do I sign?" The next thirty years of my life was now tied to a mortgage payment that I paid almost no attention to, Hillary once again overshadowing a marital rite of passage.

"Is there anything you care about more than Hillary? Anything? Jesus. And there's not even a campaign yet," Bobby said when we walked out onto the street the morning of Hillary's UN presser. He handed me the manila envelope. "Do you think you could drop this in the mailbox? It has to go today. Could you maybe squeeze that in in between Hillary since I've done everything else?"

I grabbed the envelope.

"Yes, yes, of course. I'm really sorry, but . . ."

"I know, you have to go. Just don't forget to mail it."

I tried to kiss him, but he gave me an icy cheek. I waved the envelope at him in a *you-can-count-on-me* motion. I watched his tangerine ski jacket, the one that he'd pulled from a discount rack and that every winter I tried to get him to replace with a sensible neutral color, disappear into the Midtown crowds and rushed to the far East Side.

I arrived at the UN press office hours after the rest of the horde and grabbed the first UN worker I saw to ask if I was in the right place. "I'm going to say this and try not to sound too sarcastic," she said in an upper-crust English accent. "But what about me makes you think that I care or know the answer to your question?"

The day got worse from there.

This wasn't the way Hillary wanted to start her second campaign, but there was a kind of cosmic alignment. I'd blown off a marital obligation to push my way into prototypical Clinton chaos at the UN—a bureaucratic, impenetrable organization, well intentioned but ultimately out of step with the modern era.

I waited with a delegation of Ethiopians in a security line that snaked around First Avenue. When we got through, a guard escorted us to the area known in UN parlance as the stakeout, where heads of state and special envoys talk to the diplomatic press in front of a bisque-colored wall with a UN SECURITY COUNCIL muslin backdrop. To our right was a glowing red EXIT sign and to the left were the fifteen flags of Security Council members so shoved together they resembled a tangle of table-cloths hanging at the dry cleaners. But the only visual we cared about was the almost life-size replica of Picasso's *Guernica* looming behind the podium. The press couldn't pass up the symbolism. Hillary confronting the media against the backdrop of a carnal, bloody battle scene. The mother of all advance fuckups. "I thought that painting was in fucking Madrid!" one of The Guys said.

HILLARY WAS AT the UN to mark the upcoming twentieth anniversary of her 1995 address on women's rights in Beijing. I was fascinated by Hillary's Beijing speech, both in what she said to the women's delegation ("Human rights are women's rights, and women's rights are human rights.") and in what the speech meant to her political formation. Hoping to put the failures of Hillarycare behind her and forge her own identity, Hillary, then forty-seven, defied the West Wing and insisted she address the cavernous conference hall that hosted the UN Fourth World Conference on Women. Critics at the time called the speech part of the first lady's "radical feminist agenda" and "antifamily," but even then her aides saw it, with the help of a savvy PR strategy back home, potentially becoming Hillary's "I Have a Dream" moment.

She'd hardly left the State Department when she told her team she wanted "to build on Beijing" and make known, especially to women too young to remember, that she'd been at the feminist forefront. There were plans for a Hillary Beijing Bitmoji—an animated blonde with a Betty Draper bob and powder-pink suit—and a press rollout around the anniversary. I had written a fifteen-hundred-word story that traced her career and upcoming presidential campaign back to Beijing. It never ran in print.

The email story caused all Planet Hillary's competing, concentric circles to combust.

No one frothed at the mouth blaming the email controversy on a voracious and irresponsible news media more than the Ragin' Cajun. James Carville's bald head and jutted jaw quickly blanketed the cable airwaves. "Y'all are just going to go out there and say, 'She raised more questions than she answered,'" Carville told MSNBC. Then in the least helpful defense I heard, Carville reminded viewers of the laundry list of previous Clinton scandals. "Do you remember Whitewater? Do you remember Filegate? You remember Travelgate? You remember Pardongate? You remember Benghazi?" We do now, James.

Mandy Grunwald, the mysterious media consultant long immortalized in Clinton lore dating back to 1992, suggested Hillary do a sit-down interview with someone friendly, Robin Roberts, maybe. Hillary could explain that she'd done it for convenience and a lot of the emails were about Chelsea's wedding and yoga and planning her mother Dorothy Rodham's funeral. It would all be so *relatable*.

The newcomers—Jim Margolis, the recently-hired adman; and the pollster Joel Benenson, both Obama campaign veterans—argued that Hillary should do a handful of interviews with the *Times* and the *Post*, serious outlets that most people wouldn't accuse of being too cozy. Hillary hated this idea.

In the end, she went with what she knew. The 1990s. "They won't stop until I do it," she'd said.

Confronting the hysteria head-on with a traditional press conference had worked wonders in 1994 when Hillary sat in a rose-colored sweater set for sixty-eight minutes, deftly parrying an onslaught of questions about her and Bill's investments in Arkansas. In response to a question about why she and the president hadn't handled Whitewater differently, Hillary said, "Well, shoulda, coulda, woulda. We didn't." She was cutting and sarcastic and funny. The *Washington Post* said the first lady sounded "confident and unflappable" and that the casual setting "conveyed an openness and eagerness to engage in a full give and take."

The press spent a couple of hours waiting for Hillary, crammed into the stakeout like pigs at a trough. By this point, I'd let Bobby down, possibly signed my life over to Fred from Citibank, and been belittled by a UN worker. I didn't have a lot of dignity left. I begged the official UN videographer to let me stand behind him if I promised not to move while straddling one of the legs of his tripod. I separately swore to a nearby German photographer who'd been there for three hours that I wouldn't stand all the way up and block his shot. One stray move, and we would've had a diplomatic incident.

It was in this uncomfortable scrum that the makings of Hillary's 2016 traveling press corps took shape. The Travelers included reporters for all the major wire services (the Associated Press, Reuters, Bloomberg), the newspapers that could still afford to splurge on travel (the *Washington Post*, the *Times*, the *Wall Street Journal*), the new media disruptors (*Politico*, *BuzzFeed*), and "embeds," the twenty-somethings from all the major TV networks (ABC, NBC, CBS, CNN, Fox News) who embed themselves with their assigned candidates. That day, I found myself shoulder to shoulder with the AP's Ken Thomas, who wrote upside down as he held his notebook flat against his thighs because there wasn't any space to hold the three-inch pad upright. To my right were Jennifer Epstein (aka JenEps) of Bloomberg News and Annie Karni of *Politico*, both shoved in so tight that their feet practically lifted off the carpet.

Finally, at 2:59 p.m., when the stakeout had become a stew of body odor and the German's cologne, Hillary walked to the lectern, opened a black binder, and after addressing women's rights and negotiations over the Iran deal, spoke slowly about her emails. She lifted her eyebrows and with every other sentence looked up from her prepared remarks, assembled hurriedly by her team of aides and lawyers including David Kendall and her old friend and speechwriter Jim Kennedy.

"I opted for convenience to use my personal email account, which was allowed by the State Department because I thought it would be easier to carry just one device for my work and for my personal emails instead of two," Hillary said.

She spoke deliberately, explained she hadn't broken the law in using the email account, worked in her yoga routines, wedding planning, her mother's funeral arrangements—you know, all "the other things you typ-ically find in inboxes." She explained that she "chose not to keep" those private emails. "No one wants their personal emails made public, and I think most people understand that and respect that privacy." She didn't use the D-word, but all we heard in "chose not to keep" was *DELETED*.

Hillary called on me third after a Turkish reporter (per UN protocol) who would prompt ridicule and rumors about a planted question when he innocently asked, "If you were a man today, would all this fuss be made?" MSNBC's Andrea Mitchell (per everything) got the second question.

Then Hillary pointed toward me. "Hi, right here."

For the only time during the press conference, she grinned and in an almost maternal motion batted her hand to quiet the shouts from CBS News and Al Jazeera. "She's sort of squashed, so we've got to . . ." Hillary said.

"Hi, Secretary," I said, still crouched down to avoid the German cam-eraman's wrath. Ken and I had rehearsed our questions, but I still spoke too fast, running my two-part question together. "I was wondering if you think that you made a mistake either in exclusively using your pri-vate email or in the response to the controversy around it. And, if so, what have you learned from that?"

"Well, I have to tell you that, as I said in my remarks, looking back, it would have been probably, you know, smarter to have used two devices."

Hillary could've acknowledged it was a mistake and apologized, but even as I joined the delirium of the scrum, I understood her instinct. She thought using a private email was another nonstory ("the biggest nothing burger") started by the *Times* and amplified by her political rivals. Every elected official—"even Obama," aides said (off the record)—used a per-sonal email. Why should she apologize?

One of Trump's close associates told me Trump had been baffled watching the spectacle unfold on CNN. It was one of the first times he knew that if he ran, he could beat her. Trump said he would've ended the first press conference with an emphatic "I'm DONE talking about this."

I sometimes wonder whether Hillary would've seemed more genu-ine if she'd adopted Trump's tactics. If she'd come out at the stakeout alongside a lawyer (female, fifty-something, attractive but not too attractive) and stacks of documents in manila folders displayed like props—as Trump did at one of his more theatrical press conferences. Should she have ditched the legalese and given us the *real* reason she used a private server? *Of course I didn't want the DISHONEST MEDIA reading through my emails. Look at what they've put me through for twenty-five years! That MAKES ME SMART.* Could Hillary have told my profession what she really thought of us and dismissed the email story altogether as more "FAKE NEWS from the FAILING *New York Times*"?

Well, shoulda, coulda, woulda. She didn't.

THERE WERE NO taxis so I got on a crosstown bus back to the newsroom, my laptop and MiFi balanced on my lap, my stomach growling. I typed out the first draft of the first "WHAT ABOUT YOUR EMAILS?" story. Schmidt would add his reporting to my coverage of the press conference. Carolyn would give it a read, followed by the copy desk, who scanned it for factual and grammatical errors. The slot would take a look and defy every fiber of a newspaper editor's old-school being by popping it on the home page as soon as possible. As the afternoon progressed, Schmidt and I would perfect the story, adding insights, fact-checking statements Hillary made, and making adjustments that Very Senior Editors requested until the final version for the historical printed record on page one (above the fold) was as close to flawless as was possible for a story that started on the M42 crosstown bus.

> UNITED NATIONS—Hillary Rodham Clinton revealed on Tuesday that she had deleted about half her emails from her years as secretary of state . . .

James Carville called it. The *Guernica* press conference only raised more questions than it answered. We stayed in the newsroom late into the night, studying up on government rules, talking to sources about whether Hillary would postpone her campaign announcement, scheming with Carolyn about the next day's story and the story for the day after that and the fifth-day story that could anchor the all-important Sunday *Times*.

Hillary would say later that the *Times* covered the email story "like it was Pearl Harbor." Ever since that first news conference, there was an insatiable appetite for email-related stories. I can't explain it exactly except to compare it to a fever that spread through every newsroom and made us all salivate over the tiniest morsels. Cable news talkers talked. Readers clicked, clicked, clicked. Republicans feasted. All of this heightened the fever until reporting on Hillary's emails, asking Hillary about her emails, poring over thirty-four thousand of Hillary's emails (*She got stood up for a cabinet meeting! She can't work a fax machine! She loves Upstate apples!*) became an almost out-of-body impulse turning us all into Whirlpools set on the rinse-and-repeat cycle. I never agreed with Hillary that her email server was a nonstory, especially after the FBI opened its investigation, but I would regret—and even resent—that it became the *only* story. But that was months later, when the emails swallowed everything.

In those early days, I felt invigorated and grateful to be so central in covering such a monster news event. I wanted to prove I was worthy, to myself, to my editors, to The Guys. I wanted to keep the Polar Bear alive.

I left the newsroom after 2:00 a.m. still high from closing our page-one story and from Hillary calling me "squashed" on national television. Too antsy to take a taxi, I climbed down the subway stairs at Times Square and dug around in my backpack for my MetroCard. That's when I felt the manila envelope.

13

"What Makes You
So Special?"

APRIL 2015

I couldn't sleep the night before the announcement. Bobby was visiting his family in Ireland. I wouldn't have blamed him for giving me the Hillary-or-me ultimatum after the mortgage fiasco. I lay in bed, stretching a leg over to his empty side of the mattress, wondering how many times I'd let him down over the next nineteen months. How much he'd put up with.

Not long after the email server story first broke, I made a twenty-four-hour trip to Miami for the sole purpose of shouting a "WHAT ABOUT HER EMAILS?" question at Bill Clinton as he wandered around a Liberty City housing project helping underprivileged kids during a Clinton Foundation Day of Action. I am not proud of this.

Bill wore blue jeans and a baby-blue T-shirt that hugged his paunch. He looked like he'd wandered away from a game of mah-jongg at a senior center. "I have an opinion, but I have a bias," he said when CNN asked if his wife had been treated fairly in the private server brouhaha. I pressed again. He took a step toward me. Then he stopped himself. "I shouldn't be making news on that."

Later, Clinton pulled me aside next to a bouncy castle and introduced me to a Bolivian farmer who'd relocated to eastern Arkansas. "Amy, Amy, come over here, here's a story you oughta be doin' . . ."

The Guys were always trying to get me to cover the Day of Action, but for some reason they weren't thrilled I made the trip down that day.

"I don't care what you write because no one takes you seriously," Outsider Guy told me.

I couldn't get that line out of my head. It hurt even more that it had come from him. *No one takes you seriously.*

I lay in bed studying a stack of *Times* stories I'd printed out to see how my esteemed predecessors had covered previous presidential announcements. Robin Toner, 1991: "Gov. Bill Clinton of Arkansas entered the race for the Democratic Presidential nomination today with an unstinting indictment of a decade of Republican domestic policies and a promise to restore the American dream for 'the forgotten middle class.'" The "forgotten middle class" had such lyricism and aplomb, a single evocative expression dreamed up by a centrist think tank with a sentiment similar to "Make America Great Again," minus the kitschy overreach and racist undertones. I wanted to write a sweeping lead when Hillary started her campaign.

The Brooklyn campaign headquarters had invited the Hillary press corps to an off-the-record spaghetti dinner at John Podesta's house in DC. The goal, in addition to trying to signal newly improved relations with the press, had been "setting expectations for the announcement and launch period" and "framing the HRC message and framing the race."

Podesta, Hillary's campaign chairman and a White House chief of staff to Bill Clinton and counselor to Obama, had a skeletal, hyperdisciplined frame, a distance runner swallowed by a gray suit. He exuded the cool calm and steely on-message detachment needed to wrangle the throng of friends, donors, and advisers that chased after Hillary. He'd known Hillary since the McGovern campaign in 1972, and although she wasn't as enamored with him as Bill was, they did share some endearing quirks. Podesta believed in UFOs and encouraged Hillary, also a believer and an *X-Files* fan, to promise during the campaign to "open the files" on Area 51 and to tell the *Conway Daily Sun* that "we may have been" visited already, "we don't know for sure."

Robby Mook, who had run Terry McAuliffe's campaign for governor in Virginia, would be Hillary's campaign manager. The Clintons figured that if the thirty-five-year-old clean-shaven data whiz could get Terry, the original Clinton moneyman who'd founded a fly-by-night electric car company, elected the seventy-second governor of Virginia, then he could get Hillary elected POTUS.

Robby introduced the twenty or so reporters crammed into Podesta's eat-in kitchen to the Brooklyn-based campaign team, right down to someone whose title was, I think, assistant deputy director of issues and policy. Finally, at the end of the introductions, he threw up his hands and slapped his khaki thighs and said something like "Oh yeah, and Huma! Where's Huma?"

Huma stood near a window overlooking the backyard. She gave an excruciating closed-mouth grin and waved four slim fingers at us. "Huma does the schedule and, uh, a lot of other things," Robby said. In an instant, Robby had reduced to an afterthought the most important force (whose last name wasn't Clinton) on the campaign and the person whom many of Hillary's friends would blame for her loss. Robby didn't know at the time, he couldn't have, that the "a lot of other things" Huma would do included his own job.

I'd been an insecure mess at the Podesta dinner. I was used to attending work parties with David Carr, who'd walk up to the most powerful person in the room and say, "Do you know my friend and colleague Amy Chozick?" Always putting "friend" before "colleague." In the weeks since he died, it had been hard to be in the newsroom. Editors tried to cheer me up. Dean started calling me Penguin. I didn't have the heart to tell him I was a Polar Bear.

By the time I got to the spaghetti dinner, reporters had already thrown their messenger bags and reporter's notebooks over every seat at the main table in the dining room. I ended up at the kiddie table—a four-top foldout like the one my dad set up for poker nights—squeezed into an adjacent den.

Maggie Haberman sat near the center of the main table. She didn't have to work the room. Everyone came up to her. Maggie started at the *Times* the same day as David's funeral. Carolyn's poaching her away from *Politico* was inspired. We all had whiplash refreshing our Twitter feeds trying to keep up with Maggie's reporting. I was excited to have her as a colleague, partly because you don't want to compete with Maggie and partly because she was another badass woman to join our almost entirely male politics team. I wanted us to be friends.

But The Guys used her arrival as another way to get in my head. "C'mon, you really think they'll keep you on the beat with Maggie there?" The Guys would say. And when they didn't like the way I handled a sensitive story, they'd say, "Maggie would never do this to us. . . ." I later saw emails from Brown Loafers Guy instructing donors to talk to Maggie instead of me. Their antics didn't work. Maggie was a hardened pro and even less susceptible to The Guys' bullshit than I was, but it was another attempt to undermine me and cast us on opposite sides of the Steel Cage Match. *I saw what happened to Anne. You've got a target on your back.*

In fact, the gathering was in part to signal that the era of Original Guy was over. Podesta and others called him a "cancer," bantered that he was "going off the rails," had been part of a "goat rodeo" at Hillary's State Department, and may or may not have a "disorder." OG would still pull the strings from behind the scenes, but delicately, so reporters and even Hillary's own campaign staff would think he wasn't involved.

Jennifer Palmieri, a grown-up who had held the top communications job in the Obama White House, would be Hillary's new communications director. With her deep-set hazel eyes and wispy blonde hair that softened her strict oblong jaw, Jen brought the cool breeze of sanity and grapefruit-scented bath salts to Hillary's press operation and helped me get over my phobia about The Guys. We could make small talk about shopping and weight loss, our diets and Hillary's, too. (This included Hillary staring down a mouthwatering spread of barbecue picked up on a stop at the Whole Hog Cafe in Little Rock and putting a single tomato on her plate.) Jen thought Brown Loafers Guy had potential, "if only we could beat the [OG] out of him."

Meanwhile, Outsider Guy ended up leaving his position altogether, taking a lucrative Silicon Valley job. I moved the DICKHEAD file to my archives and never again spoke to the Guy whose support, sarcasm, and frequent sharing of dog photos initially made me believe I had some inside track.

Feminine bonding aside, what set Jen up as antithetical to The Guys was that while their comebacks to reporters were like those of a rapper seizing an open-mic night, poetry in put-downs ("Is it possible to quote me yawning?"), Jen, the highest-ranking communications professional in Democratic politics, could hardly spit out a sentence. We had that in common, too. I'd listen to the audio of our conversations and transcribe quotes like "It's um, I don't, I don't, we would, uh, it is, uh, I just saw the president's, um, uh, comments, about it . . ." and "Like, we have a plan, literally."

I'd eventually see the sly genius in this potpourri of *uh*s and *like*s and *um*s, peppered with eye rolls, confused squints, and a crooked smile. Jen gave us on-the-record access while rendering virtually every conversation unquotable.

Making these precampaign receptions off the record didn't lead to candor. I almost spit out my marinara when Mandy Grunwald leaned in close to tell us Hillary would "fight for every vote" and "take nothing for granted."

During the "expectation setting" portion of another off-the-record cocktail party, this one at Joel Benenson's Upper East Side apartment, the campaign assured me Hillary's announcement would encapsulate the mood of a nation. As they framed it, her message would be nothing short of poetry, similar in scale and scope to Clinton's Forgotten Middle Class, with the historic heft of Obama promising—from the steps of the Old State Capitol in Springfield, where Abraham Lincoln started his national political career—to end "the smallness of our politics."

Instead, we got a corporate catchphrase delivered in a cheery video that might as well have been a Pottery Barn ad.

The video was posted online on a Sunday afternoon in April. In classic Hillary fashion, she'd overcorrected from her 2008 announcement video when she'd been criticized for striking an all-about-me tone, telling supporters, "I'm in—and I'm in to win," and sitting against a chintzy floral throw pillow. This time Hillary handed her video over to a montage of carefully selected nonactors, a multicultural cross section of the comfortably middle class. The black couple awaiting a first child. The earnest

Asian American college graduate applying for her first job. A gay couple strolling hand in hand talking about wedding planning. Latino brothers bursting with *orgullo* as they open a small business. Finally, one minute and thirty seconds into the roughly two-minute video, Hillary appeared standing in the front yard of her Chappaqua home to say, "Everyday Americans need a champion, and I want to be that champion."

It wasn't so much that the video was bad or did any damage. It was serviceable. It was safe. It was Hillary. But there was no meat on the bone. There was nothing for me, or anyone else, to grasp onto as "*that's* why Hillary is running for president." The best historical analogy was to Edward M. Kennedy, a front-runner ahead of the 1980 election who'd been reduced to incoherence when CBS News' Roger Mudd asked him, "Why do you want to be president?" Depending on whom you asked, Hillary had spent the past eight years or her entire life thinking about that question, yet when she started her 2016 campaign, her only clear vision of the presidency seemed to be herself in it.

EVER THE EAGER student, Hillary spent the years since the State Department poring over academic papers and dense briefing books to re-educate herself about the country she hoped to lead. She'd consulted two hundred policy advisers and grasped that sociologists and economists had concluded that Americans with annual household incomes of $35,000 to $100,000—the group any presidential candidate would need to win—no longer wanted to be called "middle class." In the aftermath of the financial crisis, for the first time since the 1960s, the middle class no longer meant the house in the suburbs, the reliable job, the family trips to Disneyland. Middle now meant underwater mortgages, college debt for kids who wouldn't find jobs anyway, dwindling retirement savings, one diagnosis away from bankruptcy. Academics started using phrases like the "near poor" or the "sandwich generation" and even revived a New Deal–era term, the "submerged middle class." But nobody was going to win the White House talking about the "near poor" or "submerged" voters.

It made me think back to a conversation from a decade earlier. I'd worked at the *Journal* for less than a year when we all stood around discussing our plans for Thanksgiving. I said I was headed back to Texas. I loved Thanksgivings in Texas, which involved turkey picked up from the Rudy's Bar-B-Q inside a gas station and boxed corn bread and a cherry-flavored Jell-O mold. One of the senior *Journal* editors said, "Texas, huh? Interesting. Be sure to check out a Walmart," as if the place we went to buy beach umbrellas and lawn chairs and paper tablecloths for backyard birthday parties was some bizarre retail concept and I was an anthropologist wading into the unknown. *Texas, huh?*

That was how I envisioned Hillary's consultants when they sat in the conference room of her private office in midtown Manhattan to contemplate what to call this curious specimen of 121 million Americans who were technically middle class. Everyday Americans. It even sounded like Walmart's "Every Day Low Prices."

THE GUYS MUST'VE told me a dozen times that the road trip to Iowa had been Hillary's idea. "I'm hitting the road to earn your vote," she'd said in the video and insisted they all hop in her pimped-out Scooby van for an incognito drive to Iowa. She always traveled with a portable radio so that she could listen to NPR—until well into the primary when her staff showed her how to listen on an iPhone. The road trip included a stop at the Chipotle in a Toledo suburb for the chicken burrito bowl heard round the world. Hillary acted as if the vacuous political press had somehow heard about the burrito bowl and rushed to cover it. In reality, Brown Loafers Guy, whose loafers and winning smile were making their way across the country alongside Hillary, tipped off a couple of reporters. This led to the leak of the grainy security-cam footage of Hillary and Huma behind dark glasses at the checkout line, *Hillary carrying her own tray.*

As far as planned spontaneity went, the Chipotle stop was a coup. "Good spin on not being noticed," Brown Loafers wrote about the news coverage. Pundits took the Chipotle stop as a sign we'd get an all-new Hillary Unbound. "We've never seen her get a burrito before," Mark Halperin observed on Morning Joe.

Hillary said she wanted to spend the early months of the campaign "getting the input of Everyday Americans." She took a cue from her 2000 Senate campaign in New York when she'd embarked on a "listening tour," sprinkling some first-lady stardust on upstate dairy farms and factory towns and bolstering her everywoman appeal.

Indeed, by the time she first ran for president in 2008, Hillary was a hands-on senator constantly in touch with her upstate constituents. That was her frame of reference during the '08 primary when the press all crammed into a living room in a prefab home in a predominantly white suburb in Indiana to see Hillary sit at the kitchen table and listen for over an hour to a proud Sheet Metal Workers Local 20 member who'd lost his job. Or when I heard her tell an unemployed waitress on the rope line in Columbus to personally follow up with her about her hysterectomy.

But by her second campaign, Hillary had spent four years traveling the world, meeting with the likes of Egyptian president Mohamed Morsi and the Burmese opposition leader Aung San Suu Kyi in Rangoon—a long way from Rochester. Hillary seemed like Rip van Winkle, awoken after a seven-year slumber to find a vastly different country. She'd missed the rise of the Tea Party. She'd missed the Occupy Wall Street movement and the rage over health care and bank bailouts and the 1 percent. She was shocked when she heard about the opioid epidemic ravaging rural communities. "This kind of snuck up on us," she remarked in New Hampshire. And when she'd learned that people no longer wanted to be called middle class—a data point that seemed a fundamental shift in the American psyche and as clear a sign as any that there was something stirring this election year—Hillary and her consultants saw only a linguistic challenge.

Perusing Hillary's paid speeches to Wall Street banks, Mandy Grunwald expressed her biggest concern. "The remarks below make it sound like HRC DOESNT think the game is rigged—only that she recognizes that the public thinks so," she said. "They are angry. She isn't."

NO ONE CALLED those early 2015 appearances a listening tour—that would've given too much credit to Mark Penn, who came up with the idea ahead of the 2000 Senate race. The events were always meticulously produced, giving them the feel of a Chuck Lorre soundstage. Hillary took her seat in the middle of a horseshoe-shaped table, or those one-sided tables only used by TV families and the Last Supper, surrounded by seven or so voters all with prescreened stories to fit the day's theme—mental health, the heroin epidemic, small businesses, affordable child-care, and so forth. The campaign instructed the Everyday Americans to call her "Hillary" rather than "Madam Secretary."

The backdrops included a packing plant outside Des Moines, where Hillary sat in front of neat boxes of fruits and vegetables labeled MADE IN THE USA. In Columbia, South Carolina, she folded her hands under her chin for a chat with minority business owners at Kiki's Chicken and Waffles restaurant. Hillary would rattle off statistics she'd learned: "The average four-year graduate in Iowa graduates with nearly thirty thousand dollars in debt." and "In New Hampshire, ninety-six percent of all busi-nesses are considered small businesses."

There was a lot of note taking and little hugging. But Hillary loved to tell everyone how much she got out of these events. "I really like listening to people," she'd say. "I learn a lot," and "I like getting out and talking and figuring out what's on people's minds." She'd call her policy team and pass on little notes she'd jotted down—like telling them to start using "opportunity system" to refer to higher education, a term Hillary learned on her stop at Kirkwood Community College. When Bryce Smith, a twenty-three-year-old bowling-alley owner, told Hillary in Norwalk, Iowa, that his biggest challenge in starting a small business was his $40,000 in student loan debt hurting his access to credit, she lit up: "I've never heard anyone so persuasively link the slowdown in busi-ness start-ups [with college debt]." Bryce got a handwritten thank-you note, and Hillary delivered on her promise to visit his bowling alley in Adel before the caucuses. (She did not bowl.)

The roundtables were entirely overrun by Hillary's staff and security detail and the press. It wasn't unusual to have a couple hundred reporters fighting for credentials to hear Hillary talk to half a dozen Iowans about community banking. Swedish, German, Japanese journalists all sprinted through rural Iowa when they spotted Hillary's van pulling up to a discussion about apprenticeship programs.

By design, there was never enough space for the press. The pen-and-papers sat cross-legged or lay on our bellies on the concrete factory floors. Cameras filled the space, and sound and lighting guys dangled their boom mics and bright bulbs over the voters' heads. This felt like a gross invasion at a children's furniture factory in Keene, New Hampshire, when a middle-aged factory worker, Pamela Livengood, broke down about her daughter's opioid addiction. "This little five-year-old lives with me, and I'm guardian," she said. Hillary gave a knowing nod. "Pam, what you just told me and what I'm hearing from a lot of different people, there is a hidden epidemic."

When the talks wrapped up, Barb, the campaign photographer, would pose everyone for photos. Hillary gave voters the same pro tip every time: "If you can see the camera, the camera can see you!"

That's when the press would shout questions. These were mostly variations on "WHAT ABOUT YOUR EMAILS?" bellowed by cable news anchors. But not always. During a discussion on small businesses at a craft brewery in Hampton, New Hampshire, a reporter yelled, "DO YOU HAVE A PERCEPTION PROBLEM?" ("I'm gonna let the Americans decide that.") After a talk about advanced manufacturing at a bike shop in Cedar Falls, Iowa, a correspondent for the *Daily Mail* yelled over and over as the half dozen or so Iowans sat dumbfounded, "SECRETARY CLINTON, WHAT MAKES YOU SO SPECIAL?" And again, the voice echoed over a question about community banking: "SECRETARY, WHAT MAKES YOU SO SPECIAL?" A *BuzzFeed* headline read Two Actual Everyday Americans Walk into a Hillary Clinton Event . . . And Get Crushed. The *New York* magazine columnist Frank Rich called the roundtables "worthy of a Christopher Guest parody."

$$14$$

The Everydays

He left behind a house in Mexico City that was neither poor nor rich, but thought itself better than both.
　—Sandra Cisneros

SPRING 2015

It didn't take long before the Hillary press corps turned Everydays into a proper noun. When we needed to get past the barricades to talk to voters, it was "C'mon, my editors need me to quote some Everydays." Or when a line of women snaked around outside Hillary's event at the Trident Technical College in North Charleston, we'd ask the campaign, "What's the crowd count on the Everydays who couldn't get inside?" And when we had to identify the participants in one of Hillary's bite-size talks about her plan for profit-sharing programs, we'd ask, "Do you have names and titles for the Everydays?" Even Brooklyn wasn't immune. When Chelsea requested a private plane to fly to an event, Podesta shrugged, "She's not an Everyday American."

But I never loved grammar more than in early May when we arrived at Rancho High School in North Las Vegas for a roundtable discussion on immigration. The plan was for Hillary to sit in the library, in front of stacks of books and guides to the Dewey decimal system, and listen sympathetically as high-achieving students told her they feared their parents would be deported. The campaign's casting department had outdone itself. There was even a transgender daughter of undocumented immigrants.

The night before, Annie Karni of *Politico* and I shared a mediocre platter of casino sushi. Bonding over our discovery that we went to the same exorbitant New York hair colorist (hers, chestnut and wispy around her face; mine, brown with honey-colored highlights that cost me a couple days' pay), we vowed to motivate each other to exercise on the road. In 2008, the campaign-trail diet, combined with hours of sitting on a bus and my complete inability to get to a hotel gym, caused me to gain twenty pounds. To lose the weight before my impending wedding, I had to "enlist" in a boot camp taught by Iraq War vets. They made me sprint up stairs carrying car tires and yelled such motivational lines as "Old people fuck faster than you run, Chozick!" I swore to myself I'd be more disciplined this time.

From that trip on, Annie surpassed the situational trail friends that any campaign reporter must make in order to endure the long hours on the road, and became an actual friend. We even spent time together when we had a rare day off the trail, drinking rosé on her Brooklyn rooftop with our very understanding husbands.

That afternoon in Vegas, she gave me a ride to Rancho High School in her bright-yellow Mustang rental. We walked in to find a troop of students hanging up blue parchment paper several times taller than they were. When they'd neatly affixed the welcome sign with white masking tape, we noticed the red letters that spelled out Hillary's campaign catchphrase, plus one glorious extraneous comma: EVERYDAY, AMERICANS NEED A CHAMPION, AND I WANT TO BE THAT CHAMPION, HILLARY CLINTON.

It wasn't just fifteen-year-olds who could see the expression didn't make any sense. Hillary knew it wasn't working. "I know she has begun to hate everyday Americans," Podesta informed the speechwriting team. But "if she doesn't say it once, people will notice and say we false-started in Iowa."

She complained to aides that the poll-tested lines they kept handing her were duds. What, she wondered, does "I want to make the middle class mean something again" mean?

But Hillary wasn't sure what she wanted to say instead. So rather than casting the Everydays to the annals of history, alongside Bob

Dole's "A Leader for America" and John Kerry's "Let America Be America Again," Hillary kept saying it and saying it. At one point, Jimmy Kimmel asked if she was jealous of Bernie's crisp "Feel the Bern" slogan "because I've seen some of yours, and they're not as good." Hillary brushed it off. "No, yeah, well, I've never been as good at slogans." But it was never about the slogan; it was about Hillary's inability to articulate why she wanted to be president. It was about Hillary, who'd been so in touch with struggling voters eight years earlier, not being able to see that Americans had to be pretty pissed off to no longer want to be called middle class.

Hillary often called herself "a proud product of the American middle class" and "the daughter of a small businessman." I heard her tell the story about her father's drapery business so many times I could mouth the word "squeegee" at the exact second it came out of her mouth. "He went down with a silk-screen and dumped the paint in and took the squeegee and kept going," she always said. (She couldn't wait to drag that squeegee all over Trump. "I can only say that I'm certainly relieved that my late father never did business with you," she preened in their first debate.)

For the rest of the spring and into the summer, donors and prominent Democrats complained to the campaign about Everyday Americans, even, in one case, raising the similarities to the Walmart slogan. Hillary told aides she felt as if she were wearing a straightjacket. In order to win over Obama's loyal voters, she needed to present herself as the center-left defender of his legacy. She said at almost every stop and particularly with black audiences, "I don't think President Obama gets the credit he deserves . . ." But Hillary also knew enough about politics to know that every election, no matter how popular the incumbent, needed to be about change. This led Brooklyn to joke that she should just borrow the slogan from the fictional fumbling Vice President Selina Meyer in HBO's *Veep*—who ran on a message of "Continuity with Change." By August the campaign started to slowly phase out the Everydays. They tested eighty-four potential replacements:

Theme: Fairness/Families
A Fair Shot and a Fair Deal.
Hillary—For Fairness. For Families.
Building a Fairer Future Today.
Fairness Worth the Fight.
Fairness First.
Putting Fairness First.
A Fair Chance for Families.
A Fair Fight for Families.
You've Earned a Fair Shot.
You've Earned a Fair Chance.
A Fair Chance to Get Ahead.
Families First.
Building a Fairer Future.
Fairness for All Our Families.

Theme: Fighter
Fighting for Fairness. Fighting for You.
She's Got Your Back.
Your Family Is Her Fight.
Your Family. Her Fight.
Your Future Is Her Fight.
Your Future. Her Fight.
A Force for Families.
No Quit.
A Fighting Chance for Families.

Theme: Basic Bargain/Making America Work
Renewing America's Promise.
Renewing Our Basic Bargain.
A New Promise for a New Time.
A Better Bargain for a Better Tomorrow.

Get Ahead. Stay Ahead.
A Better Bargain. For All.
An America That Works for You.
An America Built for You.
A New Bargain for a Stronger America.
Time for a Better Bargain.
Putting America to Work for You.
Making America Work for You.
A Promise You Can Count On.

Theme: Strength
Stronger Together.
A Stronger Tomorrow.
Strength and Fairness.
Together We're Strong.
Strength You Can Count On.
A Stronger America Working for You.
The Ideas We Need and the Strength to Deliver.
A Stronger America for a New Day.
America's Strength. America's Promise.
American Strength from American Families.
Stronger at Home.
For an America That Leads.
America Gets Strong When You Get Ahead.
A Stronger America One Family at a Time.
Strength for All Our Families.

Theme: Results/Count On
Real Fairness; Real Solutions.
New Solutions Real Results.
A New Bargain We Can Count On.
Progress for the Rest of Us.

Theme: In It Together
Progress for People.
Progress for All.
Getting Ahead Together.
Making America Work. Together.
Moving Ahead. Together.

Theme: Future/Forward
Your Future. Your Terms.
Lifting Us Up. Moving Us Forward.
Building Tomorrow's America.
Building a Better Tomorrow.
Our Families, Our Future.
Secure the Future.
A Future Worth Fighting For.
For Your Family. For America's Future.
Don't Turn Back.
Keep Moving.
Move Up.
Rise Up.
Own the Future.
Go Further.
Move Ahead.
Climb Higher.
Unleash Opportunity.

Theme: It's About You
It's About You. It's About Time.
It's About Time . . . And It's About You.
It's About You.
Because Your Time Is Now.
It's Your Turn.
It's Your Time.
Next Begins with You.

Brooklyn settled on "Fighting for Us," "Breaking Down Barriers," and "Building Ladders of Opportunity." Hillary was lukewarm on "Fighting" and "Barriers" but she *loved* the "Ladders." Months later, in the spin room after a Democratic debate, I performed the public service of telling her chief strategist Joel Benenson that the expression was a clunky garble of political talk. I could talk that way to Joel, a former New York reporter who had wisely cashed in but still liked to think of himself as a reporter. He loved to preface his regular ass whippings of reporters with "I understand, I used to be a reporter . . ." and he'd warn the campaign, "The press will love writing these. I did when I was a reporter." One editor told me that Joel got into political consulting because the *New York Times* didn't hire him. I laughed at this as I walked into Benenson's Upper East Side apartment with its winding cylindrical staircase and floor-to-ceiling bookshelves. Joel was so confident in his status as an Obama campaign alum and Clinton outsider that he'd send notes around like "We need a paradigm shift in how this world operates," referring to the "old Clinton MO."

But though Joel had some good ideas, he couldn't calibrate his neurotic irascibility to present them to Hillary with the delicacy required, as if transporting a Fabergé egg on the Amtrak.

Early on, Huma decided Joel's personality "wasn't a fit," and from then on, Hillary mostly ignored the strategic advice of her chief strategist. She did, however, offer Joel some of her own advice, telling him after the Iowa caucuses, "I've seen you on TV, you've got to remember to sit up straight." I preferred Joel's cocky combativeness to Podesta's removed nonchalance.

"Why can't she just promise to give people opportunities? Why the 'ladders of opportunity'? What's with the ladders?" I asked.

Joel went from zero to ninety in 2.5 seconds. "Well, clearly you've never needed a ladder!"

"I've needed ladders my whole life."

He shook his bearded head at me. "I don't think so. I don't think you've ever needed a ladder or you'd understand."

"Do you even know that I'm from Texas?"

"That doesn't mean you needed ladders. I'm from Queens!"

"I've needed plenty of fucking ladders, Joel. I just don't get the line."

And so on, until a journalism student whom I have nothing but hope for interrupted to ask me about internships at the *Times*. I gave him my card, turned back to Joel, who at this point was yelling, I think at MSNBC's Kasie Hunt. ("What do you mean the 'youth vote'? We're winning with voters over thirty!") I interrupted to say, "See? I just extended a ladder of opportunity." He laughed.

15

"Fucking Democrats"

IOWA, AUGUST 2015

Clear Lake has a population of 7,700 people and sits in Cerro Gordo County, a perfect square of farmland in the middle of northern Iowa, approximately twelve hundred miles on either side from the ocean. I'd flown into Minneapolis and driven three hours to Clear Lake and was circling the parking lot outside the Surf Ballroom on a Friday afternoon in August. After covering the Iowa Democratic Wing Ding dinner, I had another two-hour drive, a dull straight shot down I-35 to Des Moines. What Clear Lake lacks in geographic proximity to the ocean, it makes up for with its famous surf-themed dance hall. The ballroom, lined with plastic palm trees and built-in vinyl furniture and murals of beach scenes designed to give off "the ambience of a South Sea island," is frozen on the night of the Winter Dance Party in February 1959 when Buddy Holly, Ritchie Valens, and the Big Bopper gave their last performance before their single-engine Beechcraft Bonanza crashed into a frozen cornfield in nearby Mason City and killed everyone on board.

The speeches weren't expected to start for at least another hour, and I wasn't in a hurry to get out of the car and be tempted by the signature baked chicken wings (in spicy and mild). So I circled the parking lot a couple of times listening to the Don McLean song "American Pie." That's when the Hummer backed into me. There was the quick sound of crushed aluminum, the smell of Italian dressing from my Subway salad as it flew off the passenger seat and exploded on impact with the floor mat. The seat belt tugged at my chest so tight it nearly knocked the wind out of me. And then it was over. The Hummer inched forward.

A man in his forties, with a goatee and broad shoulders that filled out his polo shirt, stepped out of his hunk of expensive steel and testosterone. He had a pair of mirrored sunglasses perched on top of his baseball cap, and I could see the reflection of my banged-up car in them. He didn't apologize or ask if I was okay; he just slammed his car door, looked my way, and in the most visceral reminder of our divided politics, yelled, "Fucking Democrats!"

Ever since high school when my friend Kate shot the finger at a GMC pickup and the driver followed us to the Sonic and waved his shotgun at us, I knew better than to antagonize an angry man in an expensive truck. I calmly explained that I wasn't a Democrat, I was a journalist with the *New York Times*. I realized this must've sounded like a distinction without a difference to a liberal-hating Iowan, so I quickly dropped that I was from Texas.

"Your family still there?" he asked.

"Yep, my family is still in Texas," I said.

I often worked this into small talk while interviewing Everydays in the intercoastal states. It was code for *I'm not what you think I am. I shop at the Bass Pro Shop, too.* He exhaled, apologized. We exchanged emails and waited for the police. He said he owned the bar that shared a parking lot with the Surf Ballroom and invited me for a drink on the house, which was just about the last thing I wanted to do after he nearly killed me and left me with an undrivable rental car in a town 90 miles from the nearest Avis.

By the time I got inside the ballroom and arranged for a tow truck and a replacement car from Minneapolis, Hillary was taking the stage in front of a bulky American flag, and I was trying to confirm whether she'd eaten a wing. "I believe she did sample the wings," a press aide wrote. "But adding [Brown Loafers Guy] who can confirm all things wing related!" Brown Loafers referred me to a photo on Twitter of Hillary holding a Styrofoam plate of wings. "Does this mean she ate a couple?" I asked. No reply.

Hillary proceeded to string together a miraculous series of "American Pie" puns. "As the song says, 'I can still remember how that music used to make me smile' . . . and if you'll look around this room, all of you Democrats make me smile, too . . . It's clear we're ready to rock and roll."

Then, after a summer spent answering "WHAT ABOUT YOUR EMAILS?" Hillary made a joke about the server: "You may have seen that I recently launched a Snapchat account. I love it. I love it," she said, wobbling her shoulders. "Those messages disappear all by themselves."

"YOU CAN'T JOKE about this stuff, once the FBI is involved," Andrea Mitchell said on MSNBC. "Her latest dilemma is not a joke to backers who see her front-running numbers in the public opinion polls slipping," a *Chicago Tribune* column read. *USA Today* ran an editorial titled CLINTON EMAIL CONTROVERSY IS NO LAUGHING MATTER. Florida senator Bill Nelson told the Associated Press, "I don't think the campaign has handled it very well. I think the advice to her of making a joke out of it—I think that was not good advice."

Hillary was already having a bad summer. The email story wouldn't go away. She'd developed such an early and enduring aversion to us that the organizers of her speech on voting rights in Houston alerted the Travelers in advance that there would be no opportunities to ask Hillary questions. "The speech is the interview." Meanwhile, it seemed that everyone under thirty and in the entire borough of Brooklyn, where she'd leased an additional floor to make room for her expanding billion-dollar campaign, was Feeling the Bern.

Even Hillary's close girlfriends describe her as mercurial. Each morning, aides would announce Hillary's mood as if it were the weather. *Crabby with a chance of outburst* . . . Hillary isn't one of those politicians who can turn it on with ease when the cameras flash. She wore her discomfort all over her face, but especially in Iowa.

She'd been in a terrible mood when the campaign's digital team talked her into shooting a Vine video. After a brief shot of a CHILLARY CLINTON beer koozie on an iced tea set up on the banks of the Cedar River, Hillary's irritated face filled the screen, before she said with feigned goofiness, "I'm just chillin' in Cedar Rapids." The Travelers must've watched the five-second video at least a hundred times, howling louder each time we saw her windswept face overtake the screen. *I'm just chillin' in Cedar Rapids. Just chillin' in Cedar Rapids.*

Months later, Ruby Cramer—the *BuzzFeed* reporter with whom I'd bonded after one of The Guys spotted us in a sea of several hundred FOBs at a Clinton Library barbecue in Little Rock and threw us out mid-po'boy—and I would almost miss the motorcade when Hillary made a stop at Raygun, a hipster T-shirt store in Des Moines. We couldn't leave the store until we'd each bought a purple cotton T-shirt that said in white block letters I'M JUST CHILLIN' HERE IN CEDAR RAPIDS.

IN AUGUST, MAUREEN Dowd, the *New York Times* columnist and loveliest thorn in Hillary's side, published her "Joe Biden in 2016: What Would Beau Do?" column revealing that Joe Biden's dying son had urged him to run for president. I'd separately heard from Democrats and donors that Biden was seriously considering making a late entry into the race.

I had a news story in the works about Biden taking steps to run. My editor, Carolyn, lit up over the potential scoop and the ensuing capital *T*, Tension, of Biden shaking up what looked like a Hillary coronation. Carolyn had this way of making all reporters desperate to please her. Or, as my colleague Michael Barbaro put it, "Ya gotta make Mamma happy." There was nothing like the warmth of Carolyn's sun when it shined on you—her single roar of a laugh that cut through any conversation, her inquisitive eyes bursting out of their sockets upon hearing about a juicy story, the praise she heaped on her reporters when they delivered something awesome. But when she went dark—casting her light on another colleague or hardly looking away from her screen in disappointment that we hadn't brought her that killer quote or nailed down that delicious detail—life could be a cold, desperate place.

I wanted to make Mamma happy, even if it meant The Guys would destroy me for it. I'd begun not to care as much about what they thought, partly because of Carolyn's supernova support. But also because with the campaign in full swing, I had no shortage of stories to write and chatty sources informing me about the goings-on inside Brooklyn. I still wanted, more than anything, for Hillary to see me as a fair reporter, but The Guys' threats about cutting off access had no coin since I could've filled the *Times* with daily front-page stories about Hillary's altruistic work with women and girls, and she still wouldn't have given me (or any beat reporter) real access. I had nothing to lose.

I heard that Biden had confided (off the record) to the White House press corps that he wanted to run, but he added something like "You guys don't understand these people. The Clintons will try to destroy me." I'd gathered nuggets on a Draft Biden super PAC and some on-record quotes from donors, but the story was missing that final ingredient—confirmation from someone deep in the veep's innermost circle. I was at LaGuardia awaiting a delayed flight to Miami to cover a Hillary speech on Cuban-American relations (a direct aim at Jeb!) when a Biden insider tipped me off that "something big" was coming on Sunday. The only other clue he dropped was to "ask Maureen."

I fled LaGuardia and started making calls in the cab. Maureen and Biden go back to the 1988 campaign and she knows his family and friends better than most. I didn't know what she had, but I knew it would be good. As soon as her column posted, my story went up, weaving in Maureen's scoop. "Vice President Joseph R. Biden Jr. and his associates have begun to actively explore a possible presidential campaign, which would upend the Democratic field and deliver a direct threat to Hillary Rodham Clinton . . ." It led the Sunday paper.

Rather than seeing the op-ed and A1 story for what they were—a trial balloon by the Biden camp—Hillary interpreted our Biden reporting as another *Times* smear. She sent the message to all her top aides. Joel Benenson must've bitched me out for an hour. Maureen got it worse—in terms that Brooklyn would've called sexist, had they been directed at Hillary.

"What is wrong with this woman?!" Huma wrote, when a press aide emailed Maureen's column around.

"Just when she'd seemed to quiet down . . ." Robby Mook replied.

"She is full of self-loathing," Jen Palmieri chimed in.

"The front page article on Biden which was written off of only Maureen Dowd's psychotic column," Podesta wrote about my story. "They are worse than [Roger] Ailes." Days earlier Podesta had described "getting fucked by the NYT" about a story by the *Times'* Mike Schmidt and Matt Apuzzo that said the Justice Department had opened a "criminal referral" into Hillary's handling of sensitive information on her private server.

Even our own colleague, John Harwood, a *Times* contributor and CNBC correspondent, couldn't resist piling on. I can't blame him. It's like Peter Fallow, the rumpled tabloid reporter in *Bonfire of the Vanities*, says, "If you're going to live in a whorehouse, there's only one thing you can do: Be the best damn whore around."

"Cannot believe Biden story is leading the paper," Harwood wrote to Podesta. "It strikes me that 'Biden is actively exploring' is the new 'criminal referral of Hillary Clinton' if you know what I mean."

The only hitch? Both of those stories turned out to be true.

HILLARY'S CONTORTING ON issues like the Keystone XL pipeline and the Trans-Pacific Partnership (TPP) that summer had nothing to do with Bernie and everything to do with Biden. If she could secure enough support from labor and lock up the black vote, the veep would see he had no path.

When Bernie called Pete D'Alessandro to see if he would run his Iowa operation, he didn't ask whether Pete thought he could win or how much it would cost to catch up with Hillary who already had dozens of offices and hundreds of staffers spread across the state. Bernie asked only this: "Do you understand my politics?"

I kept thinking about that question. *Do you understand my politics?*

I couldn't imagine Hillary asking anyone that, and if she had asked, even her closest aides would've spent a few minutes trying to come up with the answer they thought Hillary wanted to hear.

I sympathized with Hillary. She'd maintained a consistent set of beliefs over the years, rooted in her Methodist faith and the social gospel. "A progressive who likes to get things done" was how—in an off-the-cuff moment—she defined her politics in the first Democratic debate.

But after four decades in politics she'd allowed the details to get so muddied that when she said, "Every child deserves the chance to live up to his or her God-given potential," the words felt like such a safe political platitude that reporters joked she'd soon come out in favor of kittens and rainbows.

But in those months of obsessing about Biden, while trying to keep Obama happy, she'd been forced to become a magician's assistant, stuck in a box and twisting and turning to avoid the blades. She'd stood in a canary-yellow blazer in the un-air-conditioned gymnasium of the Amherst Street Elementary School and brushed off a question about whether she supported the Keystone XL pipeline.

"This is President Obama's decision, and I'm not going to second-guess him," she said. "If it's undecided when I become president, I will answer your question."

When that didn't quite do the trick, aides spent the next couple of months debating how Hillary could come out against the project without appearing to split with Obama. "We are trying to find a good way to leak her opposition to the pipeline without her having to actually say it and give up her principled stand about not second-guessing the President in public," her chief speechwriter, Dan Schwerin, wrote to Cheryl Mills. By September, Hillary told a community forum in Des Moines that she opposed the pipeline.

Then there was the TPP, Obama's signature trade deal. At the State Department she'd called it the "gold standard" of "free, transparent, fair trade." But now that she needed organized labor to squeeze Biden out, Hillary decided the TPP didn't meet her "very high bar." Dan Schwerin said the goal of that tortured statement ("As of today I'm not in favor of what I have learned about it . . .") had been "to minimize our vulnerability to the authenticity attack and not piss off the WH any more than necessary." Mandy Grunwald wrote back, "This is so full of compliments, I can barely tell that HRC is opposing the deal."

Hillary wouldn't even take a firm stance on whether women should split the bill on a date. "Look, I think splitting the cost on a date has to be evaluated on a kind of case-by-case basis," she told *Cosmo*. "You know, many years ago I remember doing that and I know a lot of young people who even today do because they kind of consider more casual dates, group dates, to be ones where everybody pays their fair share. But I think you also have to be alert to the feelings of the person that you are dating."

16

The Ninnies

BROOKLYN, 2015
Hillary told aides that she'd "stepped in it," and not just stepped in it, but dug a foot in and rested it there so long that the stink just wouldn't wear off. She blamed her staff for the ill-advised joke at the Wing Ding. "I look forward to your feedback (Also, if anyone has a funny email/ server joke, please send it my way)," Dan Schwerin asked on an early draft.

For me, the most disappointing result of the whole thing had nothing to do with the hyperventilating media or the FBI investigation or Hillary's truly miserable mood for the next several weeks. Rather, it was that after the Wing Ding, Hillary suffered from a chronic inability to crack a simple joke. Voters never got to see Hillary's "funny, wicked, and wacky" side, as Diane Blair had once described it.

In private settings, which include closed-press fund-raisers and paid speeches to Wall Street banks, Hillary exhibited a dry, acerbic wit that didn't easily translate into lines made for mass consumption on the campaign trail. In a Q&A with Goldman Sachs CEO Lloyd Blankfein, Hillary relayed how, as secretary of state, she'd argued with a Chinese diplomat that his country had no more right to claim the South China Sea than the United States had to the Pacific.

"But they have to take New Jersey," Blankfein injected.

"No, no, no," Hillary said. "We're going to give them a red state."

But on the stump, Hillary would insert pop-culture puns that stumbled off her tongue in unexpected, often painful, ways. During the Pokémon GO craze, she told a crowd in northern Virginia, "I don't know who created Pokémon GO, but I'm trying to figure out how we get them to have Pokémon GO TO THE POLLS."

At a Jennifer Lopez concert in Miami days before the election, Hillary burst onto the stage; gave J.Lo, in her lace-up kimono and thigh-high boots, a klutzy embrace; and exclaimed to the crowd, "Let's get loud at the voting booth!"

The Travelers lived for these brief glimpses of Hillary's wacky side, spending hours trying to anticipate her next election-related wordplay. The night she joined Jay Z and Beyoncé onstage in Cleveland? "Okay, ladies, now let's get in formation . . . and VOTE." Or what about "I've got ninety-nine problems . . . and Hillary Clinton is going to fix EVERY. SINGLE. ONE."

After the Wing Ding, I understood Hillary's reflex to overthink every potential line that veered from the standard script. This, in turn, caused her already skittish aides to suck the life out of any attempt at unlitigated laughs. They fought over and rewrote jokes to the point of high parody until even the simplest punch line felt as heavy as a loaf of bread kneaded to an inch of its life by a dozen sets of overzealous hands.

The back-and-forth went something like this:

NINNY #1: I agree with Joel, let's not go back to the emails . . . What about: "I used to be obsessed with Donald Trump's hair, that was until I got to spend eleven hours staring at the top of Trey Gowdy's head."

NINNY #2: I love the joke, too. But I think HRC should stay above the [Benghazi] committee—and especially above personal insults about it. She's got every inch of the high ground right now.

NINNY #3: Agree—tempting. But she shouldn't go there tonight.

NINNY #4: Wow. You people are a bunch of ninnies.

NINNY #5: Not ninnies. We own the high ground right now. We should stay there.

NINNY #6: All that said . . . we really could use a little humor in here . . .

NINNY #7: Is there some *Apprentice* joke to make? I never saw the show. I'm also the worst person to generate jokes.

NINNY #8: You all think WJC's joke is too much about her kinda wishing after hour eight that Bernie would come through the door with his damn emails line . . . ? I think it's funny and confident and the room would love it.

NINNY #5: I think it gives Bernie the credit for putting the email crap behind us instead of her—she crushed the debate and she crushed at the committee.

NINNY #1: I defer if others think this buys us goodwill with Sanders people, but email jokes in Iowa usually end up badly and don't we want to move on?

$$\overline{17}$$

A Tale of Two Choppers

Once a traveler leaves his home, he loses almost 100 percent of his ability to control his environment.

—Agent Dale Cooper, *Twin Peaks*, 4:28 a.m. from the Great Northern Hotel

DES MOINES, AUGUST 2015

By the time I filed my story and a police report, had my car towed, got a replacement car brought from Minneapolis to Clear Lake, and drove to Des Moines after the Wing Ding, the Hilton Garden Inn had given away my room. The crush of visitors for the Iowa State Fair put my prepaid, preconfirmed suite in high demand, the late-night front-desk manager explained. He handed me a voucher for the Hotel Fort Des Moines.

"The Hotel Fort Des Moines? Isn't that place condemned?"

"It's going through some renovations," he said.

The historic hotel, built in 1919, was taller than most buildings in downtown Des Moines, erected in a redbrick H shape and framed with creepy Gothic cornices. Back in 2008, in the tenth-floor presidential suite, Bill had an apoplectic fit as the caucus results rolled in. The Clintons stayed there because it's a union hotel and they got to take over the entire penthouse floor, but it was a dump even then. There was no way Hillary or anyone who could cough up the $108 a night on the newly built Hampton Inn would choose to stay at the Fort Des Moines. Brown Loafers Guy—who had worked on the advance team that set up folding chairs and hung up American flags at Hillary's Iowa town halls during the 2008 campaign—mimed dry heaving when I told him I spent the night there. But I was exhausted, and at that hour it was the Fort Des Moines or my car.

The tawny wood-lined lobby smelled of mothballs. The woman at the front desk warned me that the air conditioning was broken and all the windows were barred closed "on account of the renovation." She handed me a plastic fan that I tucked under my arm, its cord dragging behind me. I carried my bucket to the ice machine down the smoky hallway only to find it was bone dry. I wet towels with cold water and slept naked on them.

The next morning, I took a cold shower and was out the door by seven to chase Hillary and Trump around the Iowa State Fair. The annual gathering draws one million visitors who descend on the fairgrounds for carnival rides, concerts, and the chance to talk to presidential candidates as they eat deep-fried nachos and cheese curds and seventy-five different varieties of foods impaled on a stick. The wrong culinary move can spell calamity.

In 2003, not long after John Kerry, already seen as effete, had asked for Swiss on his Philly cheesesteak instead of the usual Cheez Whiz topping, his press secretary, Robert Gibbs, spotted Kerry ordering a strawberry smoothie. He sent an all-staff email out to the campaign: "SOMEBODY GET A FUCKING CORN DOG IN HIS HAND NOW."

Hillary wore a gingham blouse and stood in front of a couple of slobbery brown cows to take questions from the press. This included several "WHAT ABOUT YOUR EMAILS?" questions ("It's not anything people talk to me about as I travel around the country," she said) and multiple inquiries about whether Tom Harkin, the lovable liberal titan who didn't endorse in 2008, had endorsed Hillary six months before the caucuses to scare off Biden. "My support for Hillary for president has nothing to do with other people," he said, following Brooklyn's instructions not to utter the veep's name.

But we knew the answer. Everything Brooklyn did the entire month of August and into September was designed to squeeze out Biden.

The press was mostly huddled under shade trees, many of us still smelling like the men's locker room at a 24 Hour Fitness. But Hillary stood fully exposed under the beating sun and didn't even glisten. It was one of her many gifts—she never sweat. I asked a close aide about this once who told me Hillary had built up a tolerance over all those years of accompanying Bill to outdoor events in Arkansas. Arkansas got hot, Texas hot, but there was no way going to the Mount Nebo Chicken Fry, the Smackover Oil Town Festival, and the Bradley County Pink Tomato Festival somehow bestowed on the governor's wife the supernatural ability not to perspire—three decades later—in August in the Midwestern humidity.

I ran all this by a *Times* colleague's partner, who is a dermatologist. He had another theory. "Botox," he said. "You get injections into your glands and just like that, no more sweat."

Under all the niceties, Iowans can be a smug, demanding bunch. They wanted Hillary to show them she could do the glad-handing, retail politicking she'd failed to do in 2008. She barreled through the fairgrounds accompanied by her usual Blob—a slow-moving, cumbersome cluster of campaign staff, Secret Service agents wearing bodyguard-casual polo shirts, at least one senator, a former governor, several prominent donors with questionable income streams, and her entire traveling press corps, at least fifty of us. "I'm just having a good time!" Hillary said.

"What did she say? We couldn't hear a word!" yelled a woman wearing an H button.

I tried to get ahead of the Blob, skipping in front of Hillary and her entourage as she made her way through the agriculture building, past the prize-winning yellow-hybrid popcorn and the display of onions and the life-size cow made entirely of butter. As she wove through the predetermined route that aides and security had mapped out for her, Hillary posed for forty-five pictures, raved about being a grandmother ("It's the best!"), examined a Monopoly-themed sculpture also made of butter ("I love it!"), asked if Iowa had any zoning programs to help out small farmers (it does), and gave an evasive "Sounds good!" to a fairgoer who asked if she would attend a tailgate party sponsored by the Iowa Corn Growers

Association. I was pretending to peruse a walk-through display on genetically engineered corn when the Blob made its way inside.

I ran right into Huma, who looked like she'd gotten lost in the Barneys shoe department. "Oh look, there's Amy," she said with her usual enthusiasm.

"Oh, hi!" Hillary said, and then turned to the corn growers. She nodded. "A lot of Americans don't appreciate how innovative and forward-looking this country's farmers are."

After ordering the requisite pork chop on a stick and taking a couple of bites, Hillary climbed into the back seat of a black SUV, leaving Iowans and the Blob behind, to fly on a private plane to Martha's Vineyard for Vernon Jordan's eightieth birthday party.

Def Leppard was performing at the fair that night, and I passed a couple fairgoers humming the grainy guitar solo of "Pour Some Sugar on Me" as I strained to see Trump's helicopter and a 1,235-pound pumpkin in the distance. The helicopter said everything about the double standard Hillary would face in 2016.

Back in 2007, the landing skids of the Bell 222 had hardly touched down on her ninety-nine-county, five-day blitz on the Hill-A-Copter before the press said Hillary's chartered chopper was the ultimate symbol of her inept campaign and her own out-of-touch elitism. "A pickup truck might have been a little more effective," snipped David Nagle, a former Democratic congressman and Cedar Falls attorney. HILLARY CLINTON COULDN'T BUY IOWANS' LOVE, a *Washington Post* headline read. I used the Hill-A-Copter in my stories, too, saying Iowans had "cringed at the loud landing of the Hill-A-Copter, which cost several thousand dollars a day in a state where voters prefer their candidates in Greyhounds."

There is a photo of us from that December morning in 2007. We are in Elkader about to step onto the red carpet laid out across the ice-covered tarmac to board the Hill-A-Copter for the inaugural one-and-a-half-hour flight to Waterloo. I can still see Hillary step out of her van with her hands in her pockets. I introduce myself and tell her I'll be covering her campaign, that we first met ten years earlier when she was in San Antonio. She pretends to remember. Hillary is wearing her black shearling-lined leather coat, the one that screams 1993 but that she'll never get rid of. I am just back from Tokyo and my hair is still a chemically streaked mass of bleached waves. I am carrying a reporter's notebook and a brown leather satchel that I splurged on before I left for Iowa because even though it is heavy and digging into my shoulder, it looks like something a campaign reporter would carry. My hair is spread all over her shoulder like some yellow sea creature's tentacles. We are so cold that she has dropped the politician look. She could be an aunt or one of my mother's friends. Cynthia McFadden, the ABC News anchor, yells from a distance, "What prepared you for this, Senator?" Hillary replies, "I wanted to be an astronaut when I was a little girl." That's when Outsider Guy, who, years later, would tell me that no one took me seriously, snaps the picture. Hillary and I are pulled in close, shoulder to shoulder, without a hint of bitterness between us. Now when I look at the photo, I realize that damn chopper hulking behind our warmest moment was an omen.

I couldn't get over how, seven years later, when Trump gave about fifty kids rides on his customized Sikorsky S-76B at the Iowa State Fair, the political class marveled at the helicopter as evidence of his marketing genius. The Trump Copter became indicative of his connection to the people. He was the Everyman's (or the Everyday's) billionaire. He told one little boy, "I am Batman"—a comment we all found *hilarious* before realizing we were living in *The Dark Knight.* One of six CNN stories about the chopper rides read DONALD TRUMP'S HELICOPTER PROVIDES A THRILL FOR CHILDREN AT THE IOWA STATE FAIR. The Bloomberg Politics headline read MARK HALPERIN RIDES IN STYLE WITH "THE DONALD." ABC News' Martha Raddatz also hopped on board. "The kids love it," Trump said over the engine's hum.

It wasn't only that a candidate had to have a penis to pull off the helicopter stunt—though that helped. It was something else, something about Trump and the press and how amusing we found all of it.

Trump stepped off a golf cart in a dark-blue blazer, his MAKE AMERICA GREAT AGAIN hat splitting the masses like the Red Sea. He'd hardly taken a step before Hillary was a distant afterthought. Trump's slow-moving cluster of press, security, and fans (I couldn't even call them supporters, not yet) holding their phones overhead, reaching out hundred-dollar bills for him to sign, dwarfed Hillary's Blob. The same Iowans whom I'd seen trying to get a shot of Hillary's blonde bob in the agriculture building now angled to get close to Trump amid shouts of "Kick Hillary's ASS!" and "Trump that bitch!"

I made an educated guess that Trump would also head for the pork chop truck and got in line, sucking in the scent of exhaust and cooking oil. I stood in line with a trio of college girls, all blondes in wind shorts that stuck to their tanned thighs. None of them were for Hillary. "I don't trust anything she says," one of the girls told me. Two of them had already decided to vote for Trump.

I could feel my forehead scrunching, my inner Valley Girl voice come out, "Seriously?" I thought for sure these girls were fucking with me or fucking with the democratic process or both.

"Yeah, I don't like everything he says, but at least he says it like it is," one of the girls said.

By the time I got back to the Hotel Fort Des Moines, it was dark outside though the humidity hadn't let up. I took the rickety elevator with its THIS WAY UP sign to the eighth floor. The door on room 817 was ajar. The smell of cigarette smoke and reefer floated through the open crack. Thinking I had the wrong room, I checked my key. Yep, room 817. I slowly pushed the door open to find a scene that looked like a meth den. Two men and one woman spread out on one of the double beds sucking down cigarettes. The mattress on the other bed had been dragged to the floor, and two more men, both in undershirts, reclined on it, their eyes little slits. I noticed a luggage rack had been rolled in with a red ice chest full of Busch Light.

One of the men on the bed, skinny with a shaved head and a snake tattoo that peeked out from his collar, took a long drag and looked me up and down.

"Can we help you?"

"Um, yeah, I think this is my room." I flashed my key as if it were a badge.

"It's our room now."

I didn't want to escalate matters. I'd watched enough *Lockup: Raw* marathons to know he'd likely graduated from a cell at the Iowa State Penitentiary to my room at the Hotel Fort Des Moines. But for the past four months, I'd squeezed everything I needed in life into a single black roller bag (carry-on size because who has time to wait for checked baggage?). I didn't see it. I'd already lost one of the diamond stud earrings Bobby had had specially made in the jewelry district and given me for our first wedding anniversary—gone forever in the sheets of a Best Western at the Manchester airport or sucked into the vacuum at the Avis return.

"Okay, well can I just get my luggage?" I asked.

"Haven't seen your luggage," the same man, the spokesman of the group, said.

"All right, well thanks anyway. I'll just check with the front desk. Enjoy your evening!" I surprised myself with how chipper this came out.

Downstairs, I banged on the bronze bell at the front desk. A petite woman with a flash of pink in her hair came out looking as if I'd woken her.

"Hi. Oh, thank God. There are people in my room and my luggage is gone and my phone died and I need to call the police and—"

She interrupted me, rubbing her eyes with the back of her wrist. "Room 817?"

"Yes. That's my room."

"Oh yeah, the Hilton only gave you one night, so I moved you out and said my friends could have the room. Let me get you your stuff." She was so polite, as if she'd given my room away to a Make-A-Wish kid, not a pile of tweakers. She turned around and came out a couple of minutes later with my roller bag.

I flung it on the floor and unzipped it. Passport, clothes, makeup, Kiehl's travel-size skin products, chargers, battery packs, backup battery packs, leather-bound journal, necklace my mom gave me before the campaign started with the bronze key that says BRAVE, ziplock bags full of almonds and melatonin—all thrown about, pulled from their neat pockets, but all there.

18

Sorry, Not Sorry

NORTH LAS VEGAS, AUGUST 2015

"Isn't leadership about taking responsibility?" Fox News' Ed Henry yelled with the grace of a livestock auctioneer.

It was August in North Las Vegas, and the thermometer in Annie Karni's rental car read 129 degrees. Hillary wore silky pants, a blouse, and pointy-toed shoes all in exactly the same apricot hue. She'd finished a town hall that hadn't gone well—nothing had gone particularly well, really, since the Wing Ding. When an ebullient supporter asked if she could get a hug, Hillary told the woman, "Sit down right there. When I finish my Q&A, I will give you that hug, I promise." We didn't see if she delivered said hug, but after the town hall, Hillary strolled to the back of the gym to take our questions.

"Look, Ed, I do take responsibility," she said. "I regret this has become such a cause célèbre, but that doesn't change the facts."

"Did you try to wipe the whole server?" Ed called out.

"I don't. I have no idea, that's why we turned it over."

"You were the official in charge of it. Did you wipe the server?"

"What? Like, with a cloth or something?" Hillary said, rubbing the air as if she were applying Windex to an invisible window.

When she stormed off leaving a glass of hardly touched iced tea behind, an NBC News reporter yelled out, "Isn't this an indication that this issue isn't going to go away for the remainder of your campaign?" She threw a dismissive wave at us and said, as she walked out an exit door and into the Las Vegas heat, "Nobody talks to me about it other than you guys."

That night the Travelers all went for a Thai feast at the Lotus of Siam, a James Beard Award–winning restaurant that shared a strip mall with a nail salon–tanning parlor combo. When the pork larb arrived (spicy level 6), one of the regular Travelers, without any malice or bias, proposed we go around the table and say whether we thought Hillary would win, starting it off with "I don't think she's going to be president. I just don't. You go next . . ."

But the rest of us just looked down at our sticky rice and papaya salad, afraid to voice our opinions even in the anonymity of an off-strip Vegas restaurant. Ken Thomas finally said that as an AP reporter he couldn't answer that question. And if the AP wouldn't do it, none of us would. As usual, I felt obligated to fill the silence, so I said something like "Well, it's hers to lose . . ."

Had I answered the question honestly, I would've said *of course* Hillary was going to win and we would all be going to Washington in a year and a half to cover the FWP. Name one Republican who could beat her. But I probably would've also said she needed to get her shit together ASAP.

I wasn't the only one who felt that way. The Clintons had planned to spend the August doldrums at a beachside estate in Amagansett that cost $110,000 for a two-week stay where they would bask in their wealthy friends' adoration while picking up $2,700 per person.

Instead, their hedge-fund manager and real estate magnate and fashion scion friends approached Bill and Hillary apoplectic about the campaign's inability to squelch the email controversy and Hillary's refusal to apologize. Bill, believing the whole thing a "witch hunt," was emphatic that she had nothing to apologize for.

Hillary blamed Brooklyn. On one conference call Hillary cut Robby Mook off: "I need a James Carville not a George Stephanopoulos defending me on this!" The analogy confounded the newer aides who didn't register the reference to the 1992 campaign. Robby was a Vermont preteen with a budding interest in school plays when the Gennifer Flowers scandal broke and everyone agreed George had wimped out.

Susan Thomases, Hillary's chief defender back then ("The Clintons'

Bad Cop," as the *Post* called her), said Stephanopoulos was "the squea-mish guy of the century—'Ooh, ooh, I can't stand this campaign,' I said, 'Then quit. Your theatrics are just adorable, but quit. If you can't handle it, quit.'"

Susan had multiple sclerosis now and was homebound. Hillary missed her. That was the kind of loyalty and defender she craved. Instead, she got a millennial armed with data and analytics.

Hillary also reminded the newer aides about her fraught dynamic with the *New York Times*. She'd seen this movie before, she told them. "They'll absolutely hammer me over emails and then they'll give me the biggest wet kiss of an endorsement and it won't matter by then," she said. (After the election, Bill would spread a more absurd *Times* conspiracy: The publisher had struck a deal with Trump that we'd destroy Hillary on her emails to help him get elected, if he kept driving traffic and boost-ing the company's stock price.)

I don't want to blame the victim. We did all hyperventilate over her emails. But Hillary and her campaign never had a strategy to change the conversation. By my tally, Brooklyn turned down forty-seven of my interview requests, including to discuss Hillary's economic proposals, immigration, her work at the Children's Defense Fund, and her years as a working mother in Arkansas. My *Times* colleagues asked for inter-views about women's issues, foreign policy, and national security, among many other substantive topics. Any of those stories would've likely kicked emails off the front page, even if only briefly. Trump understood our gluttonous short attention span better than anyone, but especially better than Hillary, whose media strategy amounted to her ignoring us.

But that August, Brooklyn knew it had to try something. When the Clintons got back to Chappaqua from the Hamptons, aides presented the Clintons with the findings from two pre–Labor Day focus groups. Voters liked when she'd shown contrition at a campaign stop in Iowa saying her use of personal email "clearly wasn't the best choice" and "I take responsibility for that decision." They wanted more of that.

Hillary mulled all this as she campaigned in New Hampshire over

Labor Day weekend. She was more introspective than usual. "When you run for office, it's a very challenging experience," Hillary said on a stop in Portsmouth.

Democrats I talked to described her campaign with a single stinging word: joyless. At the end of the press conference, *BuzzFeed*'s Ruby Cramer said to Hillary she'd wanted to ask a question that wasn't about emails but felt obligated to stick to the news.

"Oh, come on, liberate yourself and ask the question you want to ask," Hillary, ever the feminist, said. Ruby asked if she thought she was running a joyful campaign.

"I like getting out and talking and figuring out what's on people's minds," Hillary said. Then, she hoisted her arms overhead and said in a you-can-all-go-fuck-yourselves tone, "And off we go, joyfully!"

Three days later, she apologized.

It could've only happened with ABC News' David Muir. The Guys aggressively shushed me ahead of the interview when they said how flirty Hillary was with Muir and I replied I'd heard he was gay. "*Shhhhh.* Hillary can *not* know that," one of them said.

For all the lesbian theories, Hillary enjoys nothing more than flirting with a handsome, preferably straight man. She constantly talks about how she married a husband more attractive than she is, which isn't really true, but it always told me a lot about Hillary that she thought it was.

This flirty Hillary played out on the campaign trail. In Hanover, New Hampshire, she waxed nostalgic about a photographer she'd been set up with on a blind date in college, later describing him as "artistic and he was kind of poetic and in those days, you know, you were choosing between anguished young men over art and anguished young men over Vietnam . . . and he was anguished over art and all that came with it." When a beefy, smooth-headed supporter in his midforties stood up to ask a question at a town hall in Oskaloosa, Iowa, Hillary interrupted, "I do think this bald look is pretty attractive."

Even after the back-and-forth about her email server, Hillary loved to spar with Ed Henry. She would regularly look past her almost entirely female press corps to call on the Fox News correspondent, with his cherub cheeks and Pucci pocket squares.

Whenever CNN's Dan Merica, who when he grew out his strawberry-blond beard vaguely resembled a young Bill Clinton, shouted an airy question at Hillary, she'd toss her head back and giggle and say, "Oh, Dan." After Chris Christie endorsed Trump, Dan decided to ask whether Hillary was jealous that she hadn't gotten Christie's backing. Brown Loafers Guy, who'd become friends with Dan as they were among the few men on the road, said of the subsequent exchange, "She responded with a prolonged smile (you could see the gears turning) and then said 'Dan, I really like you. I really, really like you.' They are basically courting each other at this point."

Brooklyn had tested Hillary's body language with various options for anchors to conduct the apology interview. Focus groups really responded to the ease she projected sitting across from Muir, named one of *People* magazine's "Sexiest Men Alive, 2014." If she had to apologize, at least she could do it with an interviewer who didn't make her look as if she wanted to hurl across the divide and suffocate the life out of them with her bare hands, as she had looked with Andrea Mitchell and CNN's Brianna Keilar. Brianna had been selected for the first sit-down interview of the campaign in a mix-up after Huma told the press team, without specifying, "HRC wants the blonde." They hadn't been sure which blonde and chose the one who covered the campaign.

Press aides in Brooklyn buried their faces in their hands when Mitchell asked on MSNBC if Hillary wanted to apologize to the American people for her email use, and Hillary said, "I'm sorry that this has been confusing to people . . ." She gave a similar nonapology to the AP.

With Muir, Hillary finally dropped the ambiguities. "I think in retrospect, certainly, as I look back on it now, even though it was allowed, I should have used two accounts, one for personal, one for work-related emails. That was a mistake. I'm sorry about that. I take responsibility."

Later that afternoon, I'd pulled some strings from my old beat covering media to get seats at *The Ellen DeGeneres Show* three rows from the temporary stage set up above the ice-skating rink at Rockefeller Center.

Maybe it was all the talk of girl power or the adoration from the other celebrity guests—including Amy Schumer (plus roller skates), Pink, and a five-year-old presidential history buff in a miniature version of Hillary's pantsuit—but when Ellen asked about the email controversy, Hillary had already reverted. "I'm sorry for all the confusion that has ensued," she said.

19

The Pied Piper

HOT SPRINGS, ARKANSAS, AUGUST 2015
Brooklyn called him the Pied Piper—a charismatic showman who could drive the Republican rats out of the race. "We don't want to marginalize the more extreme candidates, but make them more 'Pied Piper' candidates who actually represent the mainstream of the Republican Party," read an early strategy memo to the DNC. "We need to be elevating the Pied Piper candidates [including Trump, Ben Carson, and Ted Cruz] so that they are leaders of the pack and tell the press to [take] them seriously."

An agenda for an upcoming campaign meeting sent by Robby Mook's office asked, "How do we maximize Trump?"

For the first ten months of her campaign, Hillary went around the country propping up Trump, portraying him as the unbridled id of the GOP.

After Trump called Mexican immigrants rapists and criminals, Hillary told Democrats in Little Rock, "There's nothing funny about the hate he is spewing at immigrants and their families." But she quickly added that Trump was merely saying in harsh terms what the rest of the GOP candidates believed. "The sad truth is, if you look at many of their policies, it can be hard to tell the difference."

Joel Benenson must've told me fifteen times, "They're all cut from the same cloth." Paul Begala, the affable Texan (via New Jersey) known as the yin to James Carville's yang in '92, told me Trump would help Hillary "for a variety of reasons" by exposing "the GOP as the radical right-wing anger society it has become." He argued, "The GOP's problem is not Mr. Trump; it is the extreme and angry nature of their base."

THE FIRST TIME I saw Trump he was wearing a tuxedo. He had Melania on his arm, and they wove through banquet tables decorated with white ranunculus and tea candles under the domed ceiling of the New York Public Library. I was a Condé Nast rover and an executive asked me to drop off a fat white envelope to an editor at table 2. I placed the envelope on the tablecloth and beelined for the exit before I could grab a flute of champagne or try the blini coated with caviar and crème fraîche. I was leaving the gala just as Trump arrived. There was something about seeing him that gave me the sense I'd reached a milestone of New York transplants. There's no going back to Peoria after you've been at the same party as Donald J. Trump . . . even if I was the help.

Our paths crossed several times after that, when I covered Hollywood for the *Journal* and he was promoting *The Apprentice*. Then, in 2013, Trump and Ivanka scheduled a lunch in an executive dining room at the *New York Times* to talk to business reporters about the Trump Organization. No masthead editors or political reporters signed up. Each time someone asked a substantive question about, say, P&L on the post-office hotel conversion in Washington or net revenue on licensing agreements in China, Trump would give a couple-word hyperbolic answer ("It's going to be the best . . .") and then swivel his head left to Ivanka, who would then respond in impressive granular detail, sometimes pulling slides and charts out of a camel-colored Ivanka-branded tote. Afterward, I shook Trump's hand and rushed back to my desk feeling that I'd had a nice break from cafeteria food to chat with a reality TV star and his pretty, capable daughter.

And the day Trump announced he was running, I admitted to a couple of *Times* editors that I'd watched eight seasons of *The Apprentice* and that we should do a story about it. They told me political reporters wouldn't be writing about Trump. "We have enough candidates to cover," one editor said. "Let the TV writers do it."

Months later, it got back to Trump that I'd watched his TV show. He was already the Republican nominee, and he called me to ask if I thought Arnold Schwarzenegger would be as good as he was on *The Apprentice*. "Uh, I don't think so. I don't know," I responded, cautious that anything I said could be used against me. "I don't really have time to watch anymore."

Then there was our Polish housekeeper, Wanda, who is so bad at her job that a couple of friends who've also used her for years came home early once to find Wanda watching TV in their bed. But she is a solid person and charges eighty dollars to clean a two-bedroom apartment, including laundry. She also happened to clean Don Jr.'s and Eric Trump's New York apartments. In 2014, Eric and Lara invited Wanda to their wedding at Mar-a-Lago—which, despite what anyone thinks about the Trump sons, was a class-act thing to do.

I was in Little Rock in July 2015 to cover Hillary's speech to the annual Jefferson-Jackson Democratic fundraising dinner. I hadn't seen Trump as a candidate yet, so I arrived a day early and drove an hour to Hot Springs, a quaint redneck spa town in the Ouachita Mountains where Bill spent his childhood. Trump was the keynote speaker at an Arkansas GOP dinner.

Annie and I got there early and wandered around the Hot Springs Convention Center trying to find Trump's press conference. Beautifully dressed African American families, little girls in Easter egg–pink dresses, their mothers in starling-blue lace dresses with matching hats and handbags, filled the hallways. We were so confused. "Why are *they* supporting Trump?" I asked. That's when a voice announced, "The Regional Convention of Jehovah's Witnesses is next door at the Bank of the Ozarks Arena. I repeat, *next door.*"

Trump walked into the drab conference room soaked in sweat. There were about a dozen of us, mostly local press. There was none of the commotion of a Hillary press conference. We all stood around casually ready for our Friday night entertainment.

"Ah, it's nice and cool in here," Trump said, dabbing his drenched forehead with crinkled Kleenexes he'd pulled out from his suit pocket. A couple of dandruff-size white dots stuck to his forehead and temples, like snowflakes on orange AstroTurf. I couldn't believe it. Not only did Hillary never sweat, even if she did, her team would've never allowed her to do something as bush league as wipe her sweat in front of the press. Trump flicked his wrist at a couple of photographers. "How about you go over there? You don't need to see towels on my face."

He reminded us several times that he owned "some of the greatest real estate assets in the world." He declared, "I have a great relationship with the Mexican people." And he wasted little time calling Arianna Huffington a "terrible woman" and a "major, major source of problems for a lot of people, especially her ex-husband."

I remembered crashing the Teneo holiday party earlier that year. Declan Kelly, an Irish PR man and Hillary donor whom she rewarded with a State Department position, cofounded the firm with Doug Band. They offered corporate clients access to what Band called "Bill Clinton Inc." The Guys called Declan "a fucking leprechaun."

At the party, held coincidentally at Trump's favorite haunt, the 21 Club, Declan kept talking about how Teneo "offers global solutions in a borderless world." That's when a prominent New York Democrat leaned over and whispered in my ear, "You can't sell air forever." And that's what popped in my head as I listened to Trump talk that night: *You can't sell air forever.*

I couldn't think of anything else so I asked The Donald what he thought of Hillary's plan to offer tax incentives to companies that share profits with employees. "Hillary Clinton unveiled her economic plan, part of that was profit sharing. I wanted to know what you think of her plan. Do you share profits with employees?" You can take the girl out of the *Wall Street Journal* . . .

I stopped transcribing—or really even listening—after he repeated for the third time some variation of "I've had so many great employees, many of them Hispanics." Then he shifted: "You have to understand Hillary . . ." That perked me up; after all, I'd spent most of my adult life trying to understand Hillary. "Wall Street is backing Hillary. Guys I know very well from New York, they're pouring money into her campaign. They're doing it for one reason, so don't be misled by Hillary."

Then he made a puppet motion alternating his diminutive hands up and down as if he were pulling the strings of a Hillary marionette. "She's totally controlled by people that love China." Up, down. "They will totally control Hillary just like a puppet." Up, down, up, down. Trump didn't steal Bernie's line of attack. He had it in his head all along.

A COUPLE OF weeks later, in early August, the Travelers, united by the quiet drudgery of covering the Hillary campaign, convinced her communications director, Jen Palmieri, to let us watch the first Republican primary debate with the campaign staff over pepperoni pizza at the Brooklyn HQ.

"C'mon, you can spin us in real time," I argued.

We dug into the Domino's boxes half listening as Carly Fiorina obliterated Bobby Jindal, Rick Santorum, and the other minor GOP candidates. I knew that Carly would get her fifteen minutes. A lot of what came out of her mouth was borderline crazy, but I still admired Carly's ballsy self-delusion. As Vonnegut wrote, "We are what we pretend to be . . ."

In 2010, I was writing a *WSJ. Magazine* profile on Carly's US Senate race, and we'd road-tripped up the California coast together. She'd just recovered from breast cancer and jumped into a Senate race against Barbara Boxer. She brushed off her abysmal record at H-P and even portrayed her primary opponent as a demon sheep with glowing red eyes in what remains the strangest attack ad in the pantheon of political programming. Carly and I got trashed together at the Madonna Inn, me on a house red and Carly on martinis (extra dry) in a Pepto-Bismol-pink leather booth at the Gold Rush Steak House. We then laughed our way back to our suites with their faux-brick fireplaces.

I thought maybe, despite their partisan differences, Hillary would have some feminist pride that a female candidate was standing up to

Trump while wiping the debate floor with every other GOP candidate. Nope. Hillary despised her.

"One of the reasons I'm really excited to be working on *this* campaign is because we are making history to elect the first woman president . . . Hillary Clinton," Robby Mook said when I asked if Hillary was proud that a woman stood out on the Republican side.

It's always been like that with Hillary. Feminism is conditional. "Elizabeth Dole: She <u>is</u> the woman people think Hillary is," Diane Blair wrote when Dole ran for the Republican nomination in 1999.

When the main GOP debate came on, everyone pushed their pizza crust aside and stared transfixed at the TV set up on a stand in the front of the room like substitute teacher day in middle school. During commercial breaks, Robby would pop in to talk to the Travelers. He salivated when the debate came back on and Trump started to speak. "*Shhhhh,*" Robby said, practically pressing his nose up to the TV. "I've *gahtz* to get me some Trump." Robby thought Rubio would be the nominee. Podesta was bullish on Kasich. Bill and Hillary, still stuck in the 1990s, feared the Bush surname most of all.

As the third GOP debate approached, Brooklyn assured us all that the strategy was working—Hillary had elevated Trump, using him to weaken the rest of the Republicans. Voters were starting to tune in, and the Trump sheen would soon fade. "He's a summer fling," senior aides kept telling me. It was mid-October.

20

"Spontaneity Is Embargoed Until 4:00 p.m."

No matter how hard I try, I can only see Manchester through the filter of soot that sat on the windowsill of my rented Subaru. The city is a dusky, weary, unfussy place split in two by the spiky banks of the Merrimack River. It's too far from Boston, where I'd flown into early that morning, to be part of its exurban sprawl, but too close to have the charm of rural New England. It was November, a year before the general election. I pulled the collar of my jacket tight around my neck and shivered on the short walk up the parking garage ramp and into JD's Tavern, the sports bar attached to the lobby of the Radisson where Hillary would soon arrive to share drinks, bar snacks, and some strained small talk with her traveling press.

In the seven months she'd been a candidate, Hillary had only had one off-the-record drinking session with her traveling press. I hadn't been in Iowa and missed it. But I heard a TV journalist made the rookie mistake of asking how she met Bill, allowing Hillary to filibuster with a story so immortalized in memoirs and speeches and popular Clinton lore that I could recite it verbatim. *Oh gosh, it was 1971, at the Yale Law Library . . .*

The interactions I'd had with Hillary since the campaign started had almost all happened by accident. In September, she'd stopped in at the Union Diner car in Laconia to shake hands. She met a high school French teacher who introduced Hillary to half a dozen of her students, who all wanted a photo. I stood to her left and when Hillary pivoted to make sure she'd posed with all the students, she looked at me and said, "She's not in the class. This one I know, she's not in the class. She may speak French, but she's not in the class."

SHORTLY BEFORE THAT encounter, I'd written a story looking ahead to the fall. I interviewed aides in the Brooklyn headquarters about their new "efforts to bring spontaneity to a candidacy that sometimes seemed wooden and overly cautious." The campaign brass wanted to put the Summer of Discontent behind them. The Guys came through by providing me with an interview with Robby and Jen. I'd hardly left the building when The Guys hosted a conference call with the entire press corps telling everyone almost exactly what they'd told me (minus the part where Robby joked that the high point of the campaign had been my favorable front-page profile of him). I always recorded interviews and took handwritten notes, starring anything that stood out. That way, when I listened back to the audio, I could glance at my illegible scribble and see which quotes had jumped out during the conversation and what I may not have picked up on in the moment. On the subway back to Midtown, I plugged in my earphones and listened to my Sony voice recorder, the one I'd bought eight years earlier in Shinjuku, with the Japanese writing on the side. Everything I had thought I could build a story around seemed stale now that top campaign aides had given the same talking points to the rest of the press. Toward the middle of the audio, we'd discussed how Hillary would try to show voters her softer, more personable side. It was the only original reporting I had.

That evening, I sat in Carolyn's office watching as she scrunched her forehead and lowered her reading glasses to edit my story. I could tell how rough a day she was having based on the number of empty Diet Coke cans and the font size on her screen. Carolyn cut and pasted entire paragraphs, plucking the juiciest Yorkie details (like the campaign's recognition that the Everyday Americans phrase wasn't resonating) out of a mumble of politicalese ("Our favorability is higher than any Republican"). Even as I reminded myself that a Carolyn edit always made even the dullest stories jump off the page, reading over her shoulder still gave me the reticent, disembodied feeling of a patient watching a surgeon perform an operation on their vital organs.

When she'd stitched it all back together, she shot me a "playback," what we called the edited version of a story, and sent it on to a copy editor, who debated me about whether we needed to explain what Hillary's "bowl-off" against Ellen DeGeneres meant. "I think it's obvious," I replied. The front-page headline read HILLARY TO SHOW MORE HUMOR AND HEART, AIDES SAY.

The backlash was immediate. "Today's @nytimes story on HRC read more like The Onion: Her detailed plan to show more authenticity and spontaneity. #Justdoit!" former Obama adviser David Axelrod tweeted.

The #ImWithHer crowd always assumed the campaign disliked my coverage because of HER EMAILS or my reporting on the Clinton Foundation. But that wasn't it. It was the Yorkie and Bathroomgate and that the day she declared her candidacy, I'd written that she hadn't offered a clear rationale for why she was running. Hillary hated that I'd broken the news that the mysterious "Diane Reynolds" in her private State Department emails had been Chelsea writing under her preferred pseudonym. But nothing put Hillary over the edge quite like the "Humor and Heart" story. Donors and top Democrats called her campaign to complain. Jen and Robby took most of Hillary's wrath, but The Guys got it, too, for letting me into Brooklyn in the first place. The fall, when Hillary was supposed to restart, instead became what some Democrats called the "Spontaneity is EMBARGOED until 4:00 p.m." phase of the campaign.

That's the thing about being a candidate reporter; you can't hide. If Hillary had to learn to be an inflatable bop bag, bouncing back after whatever the traveling press threw at her that day, The Guys were a cement wall, rough and unyielding and able to block me from receiving basic logistical information or asking a question. Ever since the "Humor and Heart" piece, I'd been iced out. At an event at the New Hampshire State House before driving to Hanover, Windham, and ending the day in Manchester, Hillary (and Brown Loafers) displayed a superhuman cold shoulder, looking right past my multiple shouts of "SECRETARY! SECRETARY!" Instead, Hillary answered a question about why Trump got better ratings than she did on *SNL* ("Consider the performance."), and adding insult to injury, she called on Fox News, twice.

Obama handled this dynamic differently. Back in August 2008, I wrote a *Weekend Journal* feature about presidents and body image and whether an overweight electorate could relate to Obama given his intense work-out schedule and zero percent body fat. (Not the *Onion*.) Okay, so it was an inane idea, and I was widely mocked. The headline TOO FIT TO BE PRESIDENT? and Murdoch's recently buying the paper didn't help matters.

Later that day, the press pool trailed Obama to a roadside farmers market in Florida. He ordered a strawberry milkshake, and as he took a long sip, he looked right at me. We locked eyes as he gulped down the frothy mix. Then he said, "Wow, this milkshake is delicious. Maybe if I had one of these every day, I wouldn't be such a skinny guy."

Obama then ordered strawberry milkshakes for the entire press corps.

I can safely say Hillary didn't want to buy me a milkshake that day in New Hampshire. On the upside, my reporting was right. She started to phase out "Everyday Americans." She had a new favorite line: "Get Ahead and Stay Ahead."

Asked about the email server—"I will continue doing my part for transparency. I'm also going to focus on what's most important . . . helping families get ahead and stay ahead." Pressed to release the transcripts of her Wall Street speeches—". . . I have plans that will actually help families get ahead and stay ahead." Teary stories at a town hall about the toll of the state's opioid epidemic would often be answered with ". . . we need to make sure every child can get ahead and stay ahead."

Hillary still treated New Hampshire like the womb, a safe, cozy place that made Bill the "Comeback Kid" in 1992 and resuscitated her own campaign in 2008.

Seven years later, she poured herself into the state, feeling as though she knew the place. She said she knew she'd never get the angry voters, the ones who wanted to send Wall Street bankers and CEOs and anyone with a Peloton bicycle to the guillotine. When a midlevel spokesman drafted a statement that said Hillary wouldn't "stop until everyday families can get ahead and stay ahead no matter who gets in her way," the campaign brass replied, "Too much with the 'no matter who gets in her way,' so would drop that." Hillary remained convinced the country hadn't changed *that* much while she'd been at the State Department. She was confident she could sway the convincible, rosier voters even if they disliked her. "Hey guys, be angry, and then let's roll up our sleeves and get to work," she told students at New England College. "Anger is a powerful emotion, but it's not a plan."

She told aides that during these town halls, she could see voters' posture change as she explained her practical solutions to help them "get ahead and stay ahead." Shoulders would relax, arms would unfurl, scowls would soften. She had this.

What she didn't realize at the time—and what I didn't grasp either until Bernie beat her in New Hampshire by twenty-two points—was that getting ahead doesn't mean anything to people who have nowhere to go.

I MYSELF WAS so influenced by having seen firsthand how New Hampshire had saved Hillary back in 2008 that, despite signs Bernie would win, I too believed she had a lock on the state. And she'd had such a masterful turn in the first primary debate in Vegas that the Democrats who had hankered for Biden promptly decided there was "no path" for the veep. Bernie had done Hillary a favor in declaring that people "are sick and tired of hearing about your damn emails." In contrast to her demeanor that summer, she seemed breezy and sociable that fall in New Hampshire. Which was why, after a "Fighting for Us" town hall at a high school in Windham, Brown Loafers sent an email around saying the secretary would like us to join her for "OTR drinks" in Manchester.

SHE USED TO do these things all the time. When I think about Hillary in 2008, I don't see her on the debate stage or working the rope line at a Lions Club. Instead, she is gliding between the aisles of the Hill Force One toward the press, a goblet of Yellow Tail in one hand, and balancing the other against the back of an AP reporter's seat.

In 2008, her traveling press corps was largely populated by New York guys who covered Hillary as a senator for the tabloids. The *New York Post*, *Newsday*, the *Daily News*, the *Observer*—all had reporters on the plane. (Eight years later, none of them did.) In an interview I did with Mel Brooks once, he described Dick Cavett as "astonishingly non-Jewish. He thought the Borscht Belt was something you wore." I always thought of Hillary like that, astonishingly non-Jewish. But she kept up with these mostly Jewish jokesters and would often show her sardonic side (directed, on more than one occasion, at Rudy Giuliani, and an African dictator's alleged STD) and a quick wit, even when the topics of masturbation and ferrets came up (she seemed in favor of the former, against the latter). She even revealed a romantic side (offering nut-covered chocolates on Valentine's Day and calling reporters' girlfriends—"I want to personally apologize to you that Fernando is with me and not you on Valentine's Day").

At one off-the-record drinking session with the traveling press in the wood-lined bar of the William Penn Hotel in downtown Pittsburgh, Hillary had been so candid over Blue Moon and oranges (her preferred beer) that when she left we jotted down Hillary haiku, passing around cocktail napkins and each adding a line, confessions in five to seven syllables . . .

Pittsburgh campaign stop
Blue Moon pints with oranges
I killed Vince Foster

In fact, I doubt I would have been so drawn to covering Hillary's second presidential campaign had it not been for seeing her in these settings, which had led to the dopey notion that I had unique insight into who she really was. "You just don't understand what she's like in *private*," I'd tell anyone who would listen after the 2008 primary. Another favorite observation: "Obama is charming and charismatic to fifty thousand people, but Hillary is just as charming and charismatic to five."

But that 2008 image of Hillary had faded by the fall of her second campaign. At times, we were so bored and desperate for stories that we analyzed Hillary's head gestures. "Look at that. Wow, she is nodding up a storm," John Heilemann, then of Bloomberg Politics, said on his cable show. Mark Halperin agreed, "Almost a minute's worth during not a very long event." The *Times* tallied forty-three head nods per minute in a discussion about community banking in Cedar Falls.

This created a vicious cycle.

Hillary looked at her '16 press corps and thought we were hopelessly young and driven entirely by clicks and the financial demands on our struggling news outlets. (The head-nodding stories admittedly didn't help matters.) She missed her high-minded State Department press corps, worldly, substantive reporters who didn't care about the horse race, understood the intricacies of Hillary's plan to arm the moderate Syrian rebels, and could find Kyrgyzstan on the map.

In the White House, Hillary once said that had her press corps been all women, she might do what Eleanor Roosevelt did and hold 340 press conferences. "That made a big impression on me," Hillary told reporters in 1994 after answering an hour of questions about her cattle futures at the famous "pretty in pink" press conference. "Now, Mrs. Roosevelt only invited women reporters. I don't think I could get away with that."

Twenty-two years later Hillary had an almost entirely female press corps, and she wouldn't have even known our names had it not been for the Face Book.

The campaign's press shop assembled an album with all our faces, names, and news outlets. Hillary studied the Face Book with her briefings not so that she could occasionally say, "Hi, Tamara," or "How are you today, Monica?" or "Good morning, Hannah," but so that she wouldn't mistake us for voters and—*gawd forbid*—accidentally interact with us when we approached her on the rope line.

The thing about a mostly female press corps was that Hillary likes men, preferably the damaged, witty, brilliant kind. She told aides she knew women reporters would be harder on her. We'd be jealous and catty and more spiteful than men. We'd be impervious to her flirting.

We were so starved for information that on the rare occasions when Brown Loafers did come by to brief us, we'd behave with such giddy commotion you'd think he was a Chippendales dancer at a bachelorette party. We'd swarm him, waiting to be fed. Brown Loafers would tell us (off the record, not for attribution) that Hillary planned to be in Council Bluffs and Sioux City on Friday. And we'd respond with "Thanks so much for the guidance . . ." and "Absolutely, we'll consider this off the record . . ." and "Can we do this again?" as if we'd been exclusively fed front-page headlines like HILLARY TO FILE FOR DIVORCE or HILLARY, FEARING DEFEAT, TO WITHDRAW FROM IOWA. Definitely not HILLARY TO TALK ABOUT HOW THE ECONOMY CAN WORK FOR EVERYONE IN COUNCIL BLUFFS.

I could've taken the journalistic high ground and refused to attend the OTR in Manchester, calling it a night at the Holiday Inn Express with some takeout from the Uno Pizzeria and a heaping side of righteous indignation about how cozy the rest of the Travelers were. But the truth was I wanted to cover Hillary as a real person, and to do that I needed a reminder that she was more than the tentative candidate with the tight grin whom I'd watched from a distance for months.

The carpeted dining room at JD's Tavern, up a few steps from the sunken bar area, smelled of cooking oil and malt. The room was quiet, with just a few men sipping beers and the white noise of hockey highlights on an overhead TV, when we all stormed in.

"I had that spot saved! What the fuck?"

"Did somebody move my laptop?"

"Guys, where do we think HRC will sit?"

"Where did she sit the last time in Iowa?"

"In the middle but not exactly in the middle, sort of to the left . . ."

"Your left or my left?"

"I don't know! It was a different setup."

It was like reporter roulette, everyone taking our best bet about where Hillary would sit and then banking our wide asses and puffer coats on it.

The trick had been to try to position ourselves close enough to make eye contact. The worst seats were the ones several spots away from her but on the same side of the table, making it nearly impossible for Hillary to see you as we all packed in shoulder to shoulder like schoolkids playing red rover. *Let Hillary come over . . .*

Pitchers of the local Smuttynose beer—a staple in Hillary's stump speech ever since she toured the brewery and held a roundtable discussion on small-business development with the owners ("I always drink Smuttynose when I'm in New Hampshire!")—and platters of crispy buffalo wings, flatbread, and other assorted starters sat in the center of the wooden tables for four that were pushed together to form one long, boxy table so wide we had to lift up out of our chairs and lean over our plates to grab the appetizers, dangling our press credentials in puddles of queso and marinara sauce. I got a seat toward the middle and didn't move again until Hillary had come and gone.

I don't remember seeing her come in, but from the very first—"Hi, everyone. Where should I sit?"—I could tell she would've rather been testifying before the Benghazi committee.

She took a quick survey of the two open seats, both of which happened to be to my right. She noticed me, said hello, and took the seat one away from me and next to JenEps of Bloomberg. That left the spot to my right for Brown Loafers who formed a physical barrier, in a button-down shirt and leather-trimmed messenger bag, between Hillary and me.

A martini glass appeared in front of her, placed next to an untouched pint of Smuttynose. I noticed her hands, the sunspots of a grandmother, the neat buffed nails, the modest wedding band, and the gold chain-link charm bracelet with a tiny photo of her granddaughter, Charlotte, hanging down. Seeing her scrape the cheese off her flatbread and then leave the bread on the plate, I mentioned the Whole30—a starvation diet of thirty days of no sugar, no starches, no dairy, and no alcohol that several reporters in the Hillary press corps (including me) were trying to stick to with varying degrees of success. She shook her head. Hadn't heard of it.

I can't say that she was ever impolite, but she delivered some soul-crushing (and not entirely unfounded) criticism of the political press. We were essentially gossip peddlers, uninterested in policies that affect people's lives and too dim and driven by traffic and Twitter mentions to grasp said policies even if we wanted to. At least, that's how I interpreted it. Brown Loafers would probably characterize things differently. Without our knowledge, he recorded these off-the-record sessions and sent transcripts around the campaign, the highlights of which a source shared with me.

Relations between Hillary and the press had been rocky since the start of the campaign, but that evening she exuded a particularly icy aloofness and a how-long-do-I-have-to-talk-to-you-assholes demeanor that made me feel as if I'd never been born. A younger TV reporter, less cynical than I was, later compared her disappointment in Hillary's phony response about Biden mulling a run ("You know, I didn't even think about that. I've got enough to think about.") to learning Santa Claus wasn't real. All I could think about was something I'd read in Diane Blair's journal: "HC says press has big egos and no brains."

When the conversation started to lull, a pile of discarded bread on her plate and an empty martini glass, Hillary looked over at Brown Loafers, who nodded in unspoken obedience. She patted the table with both palms and exhaled something like "Okay! Should we get going?"

THE NEXT MORNING, I woke up in my ground-floor room at the Holiday Inn Express feeling like a teenage girl just expelled from the pep squad. A tiny sliver of muted sunlight and the sound of passing cars came through the vinyl curtains. Still underneath the sheets and smelling the powdered eggs at the free breakfast buffet around the corner, I called Bobby, who was already at work and unlike me hardly ever complained. I moaned, *"She really, really hates me . . ."*

Even as I whined, full of self-pity, a part of me realized my dejection made perfect sense. Between Bathroomgate and the Biden story and "Spontaneity Is EMBARGOED Until 4:00 p.m.," why on earth would Hillary like me? And yet, for the first time in seven years I woke up clear-eyed and a little sad that ours was destined to be an impossible, tortured, and unrelentingly tense relationship weighted down by old grudges and fresh grievances. To Hillary, I was a big ego with no brain and no amount of cordial small talk could make up for the bad blood between her world and mine. "Not been good at press relations," Diane Blair wrote. "Feels intensely about zone of privacy; constantly betrayed and abused."

On the drive back to Boston on I-93 to fly back to New York, my windshield wipers smeared a layer of icy grime *right, left, right, left,* and the defroster drowned out a public radio program about interest rates. I thought of Jill Abramson, before she was ousted as executive editor, sitting on a bar stool in front of an upstairs window at an Irish pub across from the Port Authority Bus Terminal. She was offering career advice to a group of young women journalists. There were the usual concerns about work-life *blah blah blah* balance and a lament that Albany seemed impossible to cover with a D cup. "They just *assume* you're sleeping with your sources," somebody said, and all the less-endowed girls looked down at their pinot grigios. When a sheepish reporter asked how to come to terms with the distinctly female instinct to always want to be liked, while also writing the tough stories, Jill looked her in the eyes and said in her slow, deliberate way, "I always thought, you're either hated or you're irrelevant."

When I got back to the newsroom, I wrote the words "You're either hated or you're irrelevant" down in black Sharpie on a yellow sticky note that I kept taped to the center of my computer screen for the rest of the campaign.

21

"Schlonged"

Three days before Christmas Hillary Clinton gave Donald Trump a gift. The back-and-forth went like this: Trump told a Grand Rapids, Michigan, rally of presumably non-Yiddish speakers that Hillary got "schlonged" during the 2008 primaries. He also called her bathroom break during a Democratic debate in New Hampshire "disgusting." Hillary told the *Des Moines Register*, "it's not the first time he's demonstrated a penchant for sexism." Maggie Haberman and I wrote a front-page story about how Hillary hoped to use Trump's comments to energize women voters, who really should've already been energized, but that's another story.

Brooklyn mobilized its network. Democrats called Trump a sexist practitioner of "pathetic, frat-boy politics," more suited to run for "president of the fourth-grade football team." Communications director Jen Palmieri tweeted that she wouldn't respond to Trump, "but everyone who understands the humiliation this degrading language inflicts on women should. #imwithher."

Our story had just published when Trump fired a cryptic warning shot. "Hillary, when you complain about a 'penchant for sexism,' who are you referring to. I have great respect for women. BE CAREFUL!"

He didn't say Bill Clinton's name. He didn't have to. Hillary had, in Very Senior Editor's words, "poked the crazy bear that is Trump."

The strategy came right out of Roger J. Stone Jr.'s raunchy, rumor-filled book, *The Clintons' War on Women*. Scoops are not my forte. I prefer lunch-based reporting. I usually hear something newsy, let it percolate through a dessert course, and then sit on it for several days as I craft a feature around it.

But I do have a knack for convincing a range of sources to leak me books prior to publication. Seven days after Hillary declared herself "a champion for Everyday Americans," a source slid under my front door-mat a copy of *Clinton Cash*, the partisan investigation into the Clintons' wealth that Breitbart News had baked up.

A couple months before Trump's warning shot, another source emailed me (subject line "Merry Christmas"). Attached I found a PDF of Stone's 476 pages of X-rated allegations against the former first couple, including kinky sex (obviously), drugs, prostitution, and an Arkansas scheme to sell HIV-positive prisoners' blood to global plasma traffickers. "And Happy Hanukkah!" I wrote back.

Stone, a Republican operative with a head of white hair, a tattoo of Nixon's face spread across the expanse of his often-exposed back, an Upper East Side apartment packed with Le Corbusier furniture and more shoes than Imelda Marcos (we did have shoes in common), had it out for me. I heard that he'd said I did most of my reporting on Bill Clinton while on my knees, among other choice descriptions. (It was in this period that I started to develop a distant solidarity with Monica Lewinsky.)

In the book, Stone wrote, "Reporters like the *New York Times'* Amy Chozick mistakenly think that Hillary Clinton has a lock on women voters that will send her to the White House." (It wasn't lost on me that even as Hillary hated me, Stone accused me of being on the Clintons' payroll. Ah, balance.)

I wrote a brief post about the book, but my editors said Stone's allegations were too over-the-top, unproven, and conspiratorial to have any real impact on the election. They decided not to publish. I agreed.

Less than twenty-four hours after Trump's initial tweet, I was taking a rare nap. It was Christmas Eve, and I planned to cook Bobby an Irish dinner, which meant heating up premade turkey and ham, boiling pota-toes, and some Stove Top stuffing. Being Irish and married to me, Bobby didn't have high culinary standards.

I woke up an hour later, and my phone was exploding. I had hundreds of new Twitter followers and a stream of text messages. The texts came from second cousins and high school friends in Texas, from at least two senators, one sitting cabinet secretary, and a former *Wall Street Journal* colleague in Hong Kong. My first thought was that I must've inserted a terrible mistake in our story. My career was over. Then I saw @RealDonaldTrump's tweet: "Third rate reporters Amy Chozick and Maggie Haberman of the failing @nytimes are totally in the Hillary circle of bias. Think about Bill!"

Trump sat back and waited. Had Hillary responded to his threats or continued to say he had a "penchant for sexism," he may have abandoned Stone's strategy. But Hillary went silent. A close Trump aide later summed it up to me: "He knew he could throw Bill's past back at her and she couldn't say a word . . ." Who knows? Maybe he could even win (white) women.

With a couple of days before New Year's and the Bill-is-a-rapist meme ratcheting up, I emailed Trump to ask if he was an imperfect messenger, given his own very public infidelities and his ex-wife, Ivana, recently denying rumors of assault charges against him. "I believe that I am the perfect messenger," Trump replied, "because I fully understand life and all of its wrinkles."

I forwarded his response to my editors with the very professional subject line "OMG."

22

"I Am Driving Long Distances in Iowa and May Be Slower to Respond"

IOWA, JANUARY 2016

A lot was made about Hillary's predominantly female press corps, dubbed the Girls on the Bus. On any given day, in our cohort of about twenty regular Travelers, as many as eighteen of us were women. *Politico* did a story on the phenomenon. *Vogue* profiled us. This involved Irina Aleksander, an elfin Los Angeles–based freelance writer, accompanying me to an Applebee's in Polk County. The headline asked HAVE FEMALE JOURNALISTS ENDED THE BOYS-ON-THE-BUS ERA OF CAMPAIGN REPORTING?

Short answer: No. We didn't even have a bus.

For the first ten months of the campaign, I was Girl in a Rental Car. I'd spent so much time in a rental car trying to chase Hillary's motorcade while fighting the urge to check Twitter and my overflowing inbox that for my own safety and the safety of others, I set my out-of-office to read, "I am driving long distances in Iowa and may be slower to respond." On a nearly three-hour drive back to Des Moines from Monticello, with Ruby Cramer in the passenger seat and Maggie Haberman on speaker, a state trooper pulled over my silver Ford Focus. He said I was speeding . . . and swerving. As he walked to the driver's-side window and shined his flashlight inside, Ruby and I both thrust our hands up in a don't-shoot motion. "You can put your hands down, ladies, I'm not going to shoot you," he said. He entered my driver's license into the system and issued me a warning. The prospect of not speeding again in Iowa seemed pretty slim.

In New Hampshire, I'd almost driven into a frozen lake trying to keep up with the Scooby van and motorcade in a torrential downpour while navigating the state's obscene amount of seemingly unnecessary traffic circles. What did New England have against a simple stop sign? When it got so dark I couldn't see the road, I pulled into a Holiday Inn Express.

Then there was the Hummer that dinged me at the Wing Ding, an incident that made me even more uneasy behind the wheel.

Lately, each time I got behind the wheel, my mind raced. I could see the truck that would sideswipe me, its oncoming headlights, the cornfield where some poor state trooper would find my body. I even imagined how The Guys would react. Would they feel remorse, or would they say good riddance as they had when the *Rolling Stone* journalist Michael Hastings died in a single-car crash at age thirty-three? "Couldn't have happened to a nicer guy," one of The Guys said.

Then on a Tuesday in mid-January, Braden Joplin, a twenty-five-year-old from Midland, Texas, with a newly grown beard and tender brown eyes, died at 4:30 p.m. CST in the Nebraska Medical Center's trauma unit. I read all the details. The driver lost control of the van that was carrying Joplin and three other Ben Carson campaign volunteers between events on an icy stretch of I-80 near Atlantic, Iowa, causing the van to careen over the median and collide with an oncoming Chevrolet Avalanche. I urged the campaign to get us a press bus.

In 2008, the Travelers all rode together in a press bus paid for by our news outlets and coordinated by Jamie, the Clinton campaign press wrangler whom we all came to adore. The 2008 crew so settled into our bus that we'd each mark our territory with power cords and raggedy scarves. Any newcomers outside our established cliques would be destined for the Landfill Seats—what we called the back row of seats sandwiched between the trash and the toilet.

But eight years later, Robby didn't want to spend money on a Jamie-like staffer. At least that's what the campaign told us.

From the start Hired Gun Guy told us, as if in a studied trance, the same thing about Robby. That Robby, *what a tightwad.* Man, he is *cheap.* He's such a *bean counter.* There were carefully placed stories about Podesta riding a thirty-dollar Vamoose bus between DC and New York.

Robby's cheapness was supposed to signal to donors that there would be none of the lavish flourishes of Hillary's '08 effort, which ended so in debt that she had to loan her campaign $13 million of her own money—an experience that made her a real coupon-clipper in '16. There would be no chartered Hill-A-Copter. No $95,000 spent on sandwich platters for caucus-night parties when there was nothing to celebrate. No splurging $3,000 on six hundred snow shovels for elderly caucus goers when the forecast didn't call for snow (and any self-respecting Iowan already owned their own snow shovels, thank you very much, New Yorkers).

But, once again, Hillary's biggest missteps of 2016 stemmed from trying to prove she'd learned from mistakes made in 2008. Robby's penny pinching meant organizers and volunteers in Ohio and Florida had to go to Walmart to buy clipboards and pens. I talked to one organizer in North Carolina who'd rolled his mom's office chair down the sidewalk to man a volunteer sign-up table because the campaign didn't provide chairs. Supporters begged for lawn signs and bumper stickers. Most never got them.

Many months later, during the general election, Brooklyn resisted dispatching resources to Michigan and Wisconsin, despite on-the-ground pleas from labor unions that Trump was gaining there. Ed Rendell, the cocksure former Pennsylvania governor who has butt-dialed me more than once, urged Brooklyn to spend more to reach suburban and rural parts of his state, but he was always told no. Three weeks before the election, Brooklyn stopped polling altogether in Pennsylvania, Florida, and other states. By one estimate, the campaign ended with $20 million in unused funds, which could've paid for a pile of yard signs and polling and targeted ads in the Rust Belt but instead partly went to Jill Stein's recount efforts in Wisconsin.

After Braden's death, I pleaded with the campaign on behalf of all the

Travelers, "We all worry it will soon be untenable and unsafe to travel between events."

With no commitment on the bus—and Hillary content to have us trailing her at eighty miles per hour like paparazzi—I landed in the Quad Cities bracing to spend the final month before the caucuses driving on Iowa's frozen farm roads. I asked for the biggest, baddest motherfucking truck they had. The black Toyota 4Runner had stained upholstery and no power anything but looked as if it could've been driven by a dude in a monster truck show crushing all my months of Ford Focuses in its wake.

The woman at the Avis counter apologized that the layer of grime from off-roading couldn't be washed off because of the freezing temperatures. I examined the truck, dangling its keys around my index finger. It had a don't-fuck-with-me silver grill, half-a-foot lift on the enormous all-season tires, and by some stroke of rental-car fate, Texas plates.

23

Meeting Our Waterloo

DES MOINES, JANUARY 23, 2016

"And what rough bureau, its hour come at last / Slouches towards Des Moines to be born?" John, the news assistant, emailed the politics team. The *New York Times* had arrived in Des Moines.

Our temporary politics bureau was up and running in the Waterloo conference room, a windowless space with loud carpet, beige walls, and a low-hanging popcorn ceiling on the third floor of the Des Moines Marriott near a covered walkway that led to a Quiznos and the parking garage. I'd parked Beast, my name for the 4Runner, and walked inside to see the sign that displayed the daily forecast (HIGH 32 / LOW 17), a stock image of a martini resting on a bar, and WE'VE BEEN EXPECTING YOU: NEW YORK TIMES NEWSROOM.

In the two years, six months, and twenty-four days I'd been on the beat, I had the freedom to travel almost entirely without adult supervision. I once went to Montana just to try to smooth things over with one of The Guys during a dinner of artisanal beer and cow testicles. I'd been on so many trips to Little Rock that I'd turned one of the doormen at the Capital Hotel into a source ("Pile of Clinton old-timers here. Something's up," he'd whisper at check-in). I knew where to find homemade tamales in Vegas and a surf-simulation workout in St. Louis. I'd racked up so many Avis miles that when Bobby and I rented a compact car to drive from Miami to Key West, they welcomed us on a red carpet and presented me with the keys to a convertible BMW Z8 roadster.

The travel came at a cost. I was lonely. I missed my work wife, Michael Barbaro, and his daily commentary on my wardrobe ("I would just maybe wrap the scarf this way . . ."). It was like a tree falling in the forest—was it even a cute outfit if Michael didn't see it?

New York magazine had deemed Michael—with his round-framed glasses, salt-and-pepper curls, and a goatee so groomed that its mustache and slender chin line practically formed a *Times*ian *T* split up the center of his cylindrical face—a member of the *Times'* "Gay Mafia" of political reporters. He had a Yale degree, an Upper West Side co-op apartment, and he took wide, confident strides around the newsroom, all of which camouflaged that he had as big a chip on his shoulder as the rest of us. When I first met Michael, I only saw his posh exterior. But by the time we occupied adjacent cubicles in the politics pod, gossiping and cracking that's-what-she-said jokes all day over our shared partition, I knew him as the son of a New Haven firefighter who had scrapped to get his foot in the door. In addition to fashion critiques, Michael became an invaluable resource as I tried to keep up with interoffice gossip while on the road.

I did everything I could to both travel *and* deliver stories. I wanted to think that at heart I was still a foreign correspondent, whether it was a coffee shop in Cedar Rapids or a factory in Shenzhen. But in the era of live streaming and Twitter and a candidate who hardly acknowledged us and almost never broke script, my editors often didn't see the point. I was about to take my seat on a JetBlue flight back to New York from Orlando when one editor said, "I worry you'll be in the air and we need this for the front . . ." I ran upstream, pushed past passengers as they loaded their carry-ons overhead, and demanded that the flight attendant let me off the plane before takeoff.

"You know you can't reboard?" she said.

"Yeah, got it," I said.

I then sat on the floor of the Orlando airport near a power outlet to write the story while industrial carpet cleaners passed by. I ate a stale Starbucks bistro box and checked into the Hampton Inn. The next morning, I got the first flight back to New York and bought the paper with my A1 byline.

Adding to my urge to travel, I knew that even as Hillary ignored me, it pissed her off when I skipped her events. After I'd taken a couple of weeks off the trail, I ran into Huma at the annual Blue Jamboree fundraiser in North Charleston, South Carolina. I said hello as she stood under trees dripping with Spanish moss. "Oh, Amy," she said behind her dark cat-eye sunglasses. "Are you still covering this campaign?"

Just as my stint in Tokyo had, my months traveling the country started to create a schism between my new life on the road and my old one back in New York. I began to feel more at home among strangers in the TSA Pre line than with Midtown office workers shouting out their coffee orders. On days when *Politico* scooped me or Twitter directed its ire my way—in other words, most days—I took comfort in knowing that no one lined up at the Cinnabon at O'Hare cared what was happening in my little world.

I'd find myself at a rare Sunday brunch back in New York cringing at the conversations I'd overhear, about summer shares on Fire Island and gallery openings and seeing *Hamilton* for the second time "with the new cast," things I would've talked about, too, in my old life. But that was before I spent most of my days talking to people in the middle of the country who couldn't afford health care or lost their jobs when the factory closed or took out a second mortgage to pay their kid's college tuition. Everything back in New York started to seem so petty.

And each time my editors summoned me back to New York, after a couple of days making calls and watching the election unfold via Twitter and CNN, I would plead to go back out.

This election isn't happening in my cubicle.

Pat Healy, who in addition to Anne Kornblut was the only other person who understood what it was like to have your personal life and psychological well-being upended by Hillary and her cadre of adoring men, had also been in Des Moines for weeks. For years OG, echoed by his minions, had told me Pat wouldn't hesitate to take me out. The opposite happened. Pat's arrival on the politics team made me feel as though the target on my back had faded, or at least been covered up under so many layers of fleece and down that I didn't walk around with it like a scarlet A anymore. Pat was the most fair-minded reporter I knew.

If The Guys had tried to take him out, too, then maybe David Carr had been right and it wasn't me.

Pat and I both arrived in Iowa several weeks before the caucuses. He missed his boyfriend, Ray, and I missed Bobby. We made a pact to comfort each other with daily over-the-top, mildly sexual compliments. "You look so hot in that down vest," he said when I climbed into the booth across from him after three hours of sleep and reached for the bottomless coffee. Pat wore sweatpants and a white undershirt. He is objectively hunky, with fervid blue eyes, a head of black waves, and a few days of stubble.

"Have you been working out? Your chest looks ripped," I said. And then we'd giggle and carbo-load on Lender's bagels and cheese Danish at the buffet at the Rock River Grill & Tavern inside the Marriott because we never knew when our next meal would come.

Then, in the final week before the caucuses, the person I'd been in New York and the person I was becoming on the road, two sides of myself that no longer seemed to fit, collided. Carolyn, a half dozen editors, a handful of support staff, and at least twenty political reporters all descended on Des Moines.

They seemed so out of context in Iowa, so blatantly not of the place, like when my parents visited from Texas and I took them for avocado toast on the Lower East Side.

"Hey! Welcome to Iowa. Wow, everyone is here . . ." I said as I burst into the Waterloo to greet my colleagues. I quickly realized that Midwestern friendly didn't quite fit the vibe.

A couple of people mumbled "hi" without looking up from their screens. A social media editor gnawed on her fingernails. The web producer next to me quivered his right leg so wildly that I felt the folding table vibrate.

Descriptions of our proposed stories were splayed on a whiteboard, and I watched Carolyn stand up and with the flick of her wrist kill several slugs. By then, the 2016 story was moving so fast that the features and analysis pieces we'd proposed, or that editors had ordered up, could by the end of a day's news cycle already become stale. My heart ached as I watched UNIONJEB, followed by BERNIEKIDS and DEMSFRACTURE, swept from the annals of history with a single eraser swoop.

The Waterloo had the silent, harried feel of the final minutes of the SAT had the test been administered in a Siberian labor camp. "What the fuck?" I mouthed to a colleague who only shook his head and drooped his eyes back down to his keyboard. "Mommy and Daddy aren't speaking right now," he G-chatted me, which was what we said when Carolyn and her top deputy stewed after a heated debate.

The arrival of the *Times* in Des Moines signified more than my own reining in. The race had finally become real. For the first time, people would vote or caucus (whatever that means) and the story we'd spent the past couple of years preparing to cover would take on its own wild, unpredictable life.

This led to our descent from steady and manageable low-grade hysteria to such high-grade hysteria that when Hillary said we need a more open conversation about mental health care, my first thought went to the *Times'* politics team. My editors implemented a 7:45 a.m. ET daily conference call during which, in a hungover haze, we'd all try to one-up each other with our Definitive Hot Takes. "Cruz has sealed up the evangelicals . . ." and "Hillary's gonna hit Bernie on foreign policy today . . ." and "Trump's debate scam won't work." Carolyn, aside from a couple of grunts and "Okay, who's next?" hardly ever reacted. That made us all even more hungry.

I hadn't seen Bobby in weeks. In Japan, we'd always made time to talk, but on the trail, I hardly had time to call to say goodnight. He'd booked a flight to come visit that weekend. He'd heard so much about Iowa and wanted to see it for himself in all its flat, bland American glory. I adored that Bobby wanted to experience "life on the trail," as he called it.

There would've been no trail for me without him encouraging me to stick with it, and he somehow didn't hate the road for keeping us apart. He met up with me whenever he could. With Bobby there, even the simplest trail activity, heating up Hot Pockets in the lobby of a Hilton Garden Inn in Manchester, became an adventure.

"Life on the trail babes . . ." he'd say, pulling down the floral duvet and spreading a picnic of junk food out on the sheets of one of the room's double beds.

In 2008, I had even snuck Bobby into a primary debate. He held my reporter's notebook and told the organizers in his thickest Irish accent that he was with *Horse & Hound* magazine.

He'd already planned to drive to Iowa City and visit a science center in Des Moines so that I could work. But sensing Carolyn's stress level, I worried she and my other editors would see Bobby in the hallways at the Marriott and frown on my trying to squeeze in personal life a week before the caucuses. I told him to cancel. The line went quiet. I sensed his tolerance for my travels, my selfishness, my always putting Hillary and my job before everything else, wearing thin.

"It's too late to get a refund," he said. "I haven't seen you in almost a month."

"I know, and I'm dying to see you, but I just can't right now. I'll pay you back for the flight," I said, an empty promise since our money all came from the same place. We hung up further apart than we'd been in years.

I was pretty focused and stoic about another week without Bobby until my Glow app sent me the super-helpful pastel-purple alert, "Your fertile window is closing . . ."

This led to my sobbing to Maggie Haberman over a chardonnay in the hotel lobby. She looked up from Twitter, over her glasses. "Okay, you need to take that app off your phone immediately," Mags said.

24

The Girls on the Bus

IOWA, 2016

"To our traveling press corps—Happy New Year!" the email read. "For your safety and convenience we will be providing a bus that will begin in Davenport and transport press throughout the swing."

I had an excuse not to work from the Waterloo. We'd finally gotten our press bus—a glorious maroon Signature premium people carrier with TVs over every third row and boxed lunches and bottled water piled up on the front couple of rows and power outlets under all our seats.

For the Travelers, the arrival of the bus—parked in all her pack-journalism splendor on the frozen Mississippi Valley Fairgrounds in Davenport—signified more than an end to speeding tickets and Avis points. We'd finally moved into our very own communal home, like a loft apartment on MTV's *Real World* but with wheels. In the outside world, most of us wouldn't have chosen to spend our time together and certainly not *that* much time together. But in our shared caravan, we were the Travelers. The bus marked the beginning of us becoming a rowdy, high-strung family forever bound by our bizarre lifestyles, unhealthy diets, and constant search for a power outlet.

The nine or so of us on that first bus trip wanted to mark the moment. We stood on our seats and squatted in the aisle to fit into a group photo. "Say 'I'm With Her!'" a young campaign staffer said. "Can you just take the picture?" one of the Wires said.

Like all candidate reporters, I'd devoured Timothy Crouse and Hunter S. Thompson and Richard Ben Cramer and David Foster Wallace's *Up, Simba!* (plus glossary), romanticizing the campaign bus beyond all reason. I imagined Great Men, the "heavies" as Crouse called the top rung on the hierarchy of traveling press, Johnny Apple (*NYT*), David Broder (*WashPost*), and Bob Novak (*Chicago Sun-Times*), driving public opinion in between drinking sessions. Their prose had the power to sway primaries and make other Great Men into presidents or tear them down until they were also-rans confined to a historical footnote (see Muskie, Edmund). The job had a poetic, renegade feel. Men left their wives and families and their comfortable homes in the suburbs to sleep in a different hotel every night. All in the service of Democracy and dick swinging. Add to the political clout free-flowing booze and summer-camp camaraderie between reporters who spent our days together in buses and our nights at the hotel bar on an expense account, and well, it was hard to believe that anyone got paid to have that much fun.

That's not to say I felt like much of a journalist on the campaign trail. Writing an entire story on deadline on your lap from a swerving press van in a motorcade while trying not to spill what's left of a four-hour-old venti cold brew does require a certain skill set, but it never felt quite like journalism. Even in '72, the reporters assigned to travel everywhere a candidate went didn't exhibit much intrepidness. The Travelers knew the candidate and the campaign better than anyone, but we also had to grudgingly accept that we existed mostly for protective purposes, "in case something happened."

The '08 campaign had hardly been a media lovefest. The Travelers had been so pissed at Hillary for ignoring us at one point that when she made a one-minute-twenty-eight-second visit to the press bus offering bagels and coffee ("I didn't want you to feel deprived") no one partook. Reporters actually turned down food. *That* was how bad it was.

Hillary had to apologize after one of the '08 Guys nearly blew a gasket on us before the press bus even pulled out from the Marriott. "You know, he's not the person I would have put in charge," she said. Another '08 Guy told the bus driver to speed up when he spotted ABC News' Kate Snow sprinting through the parking lot of a Hy-Vee in Des Moines after she'd lingered too long at a meet and greet with Bill, Hillary, and Magic Johnson. The campaign once put the traveling press file in a men's locker room, where I wrote my story while sitting between Tina Brown, the glamorous former *Vanity Fair* editor, and a urinal. ("These accommodations should in no way be taken as a comment on the quality of our media coverage," the '08 Guys said.)

But at least back then the "traveling press secretary" actually traveled *with* the press. This didn't seem like a radical concept until 2016 when on most days not a single person authorized to speak for the campaign ever traveled with the press. Proximity was power in 2016. They preferred instead to ride alongside Hillary in the motorcade or on her private chartered plane.

On one swing, when there was no room on the charter between Iowa City and Ottumwa, Brown Loafers sat on the plane's turned-down toilet seat for the half-hour flight rather than ride in our putrid press quarters. On a typical day, we'd spend eighteen hours on the bus only to set eyes on Hillary from the back of a packed gymnasium or as a flash of blonde disappearing behind a van door held open by a bulky Secret Service agent.

I THINK IT was Cheryl Mills who said that "by the time women and minorities reach the presidency, the role has been vastly diminished." Well, call it a slap from the patriarchy or a stroke of bad luck, but by the time women reporters dominated Hillary's press corps, Twitter and live streaming and a (female) candidate who had zero interest in having a relationship with the press vastly diminished the campaign bus's place in the media ecosystem.

My colleagues could cover a speech or a press conference (on the rare occasion those happened) while watching the live stream from New York where they'd have Wi-Fi and power and wouldn't have to worry about waiting in line at a porta-potty on deadline or some fresh-faced campaign staffer yelling "LOADING!" right when you're crafting the perfect nut graph.

The traveling press had become the province of what one prickly print reporter (on his way to a buyout) called "the Human Tripods," the young network embeds who'd never covered a campaign before and who had to capture everything the candidate did on video. As long as the Tripods delivered a live stream, the print reporters could do our jobs and the ecosystem worked. The *Times* and the *Post* and the AP and *Politico* still broke news and provided TV talking heads with something to gab about. But in our little leper colony on wheels, the masters of Snapchat and Vine and Twitter and Periscope had become the new "heavies."

"He practically goes around with a T-shirt saying 'I work for the *Times*: I'm Number One!'" That was how one politico described the *Times*' Johnny Apple to Crouse during the '72 campaign. I approached our inaugural bus outing, a two-day swing to Davenport, Cedar Rapids, Des Moines, Osage, Sioux City, and Council Bluffs, ready to project a similar haughtiness. I even wore a synthetic wool hat that I'd bought at the company holiday sale, with a stretched THE NEW YORK TIMES in white embroidery across my forehead and that almost everyone made relentless fun of. *I work for the* Times: *I'm Number One!*

My sense of superiority lasted about two hours and thirty-five minutes. That's when I found myself somewhere on I-80 perched over the back of my seat pleading with the embeds to let me watch their video feed of Hillary's town hall. Because Hillary preferred to fly to her events (and really, who wouldn't?) the bus-bound Travelers couldn't make it to the Cedar Rapids and Osage stops. Our only option was to live-stream Hillary's Iowa events from the press bus in Iowa.

Then, through a muffled intercom, the driver, whose name tag read CHUCK, WEST DES MOINES, IA, apologized. All I heard was, "So sorry folks . . . gotta . . . generator . . . break . . ." The power and the Wi-Fi went out. We could live without Krispy Kreme donut holes and Chips Ahoy! snack packs. We could even hold our noses over the toilet that had several counties back run out of antibacterial hand foam. But the prospect of losing Wi-Fi as Hillary carried on without us in Cedar Rapids pushed the Travelers over the edge. How would we explain to our editors that we'd allowed ourselves to be sequestered hundreds of miles from the candidate we were supposed to babysit? I imagined something terrible happening—a terrorist attack or an assassination attempt. My editors would pull me off the trail forever. I could hear the scorn: "You had ONE FUCKING JOB!"

That's when the world's most influential print publications—the *Times*, the *Post*, the *Journal*, *Politico*, the AP, Bloomberg, and Reuters—banded together and did the one thing we still felt empowered to do. We whined . . .

"She could've been shot!"

"Yeah, or dropped dead of a heart attack."

"Seriously, guys, what if something happened to her and we weren't there?"

"The bus fucking sucks."

"I HATE the bus . . ."

"Sorry, Chuck."

"How much longer?"

THE DAYS CONTINUED like that . . .

1/23: Davenport → Clinton → Davenport

1/24: Davenport → Marion → North Liberty → West Des Moines
    → Davenport

1/25: Des Moines → Waukee → Knoxville → Oskaloosa → Des
    Moines

. . . until the campaign had become one endless bus ride through frozen cornfields.

My whole body and my journalism atrophied on the bus. On most days, I'd make at least a dozen calls to sources, but on the bus I hardly made any phone calls or talked to anyone outside my fellow Travelers. I no longer had the energy to yell at my editors when Pat Healy or Jonathan Martin got to write the daily A1 stories. I lost my will to protest when editors only wanted me to send color and quotes that would be melded (or not) into the roundup Frankenstory, what we called the editor-assembled daily news stories with multiple bylines and several contributor lines at the bottom. I didn't even complain when the Travelers had to convene in the lobby of the Marriott at 7:00 a.m. only to drive to the Jewish Federation of Greater Des Moines and sit on our bus outside as Hillary answered questions about Israel. The campaign said the space was too tight to accommodate her largely Semitic traveling press corps. During the event, Hillary had a mild coughing attack, or at least it looked as though she'd had a mild coughing attack from the live feed that I watched on my phone while standing in the parking lot puffing on an e-cigarette. I hadn't smoked since high school, but at thirty-seven it seemed like as good a time as any to develop a nicotine addiction.

The Subway sandwich situation had become so dire that I gave some poor freelancer a lecture about bus etiquette after he'd grabbed two turkey sandwiches, leaving me to eat the shredded lettuce, pickles, and tomatoes off a foot long that I'm pretty sure had fallen on the floor.

We started to hoard food. I once filled a Hefty garbage bag with hard-boiled eggs, hummus packs, and fruit trays from an Au Bon Pain in Indianola and slung it over my shoulder hobo-style as we trooped on and off the bus. Irina, the *Vogue* writer, who was Russian, watched me do this and asked if I'd ever been an orphan in the Soviet Union.

We reverted to tweens. The bus almost abandoned us in Vinton (pop. 5,257) after we couldn't pull ourselves away from *The Fast and the Furious* arcade game at the roller-skating rink where Hillary spoke. She declared, "The entire country, indeed the entire world is watching to see what happens right here in Benton County . . ." The entire world except the members of her traveling press, who were in the adjacent room locked in a heated game of Ms. Pac-Man. We established cliques, banishing newcomers to the Landfill. We started our periods at the same time and sang Justin Bieber's "Love Yourself" on a loop.

WHILE OUR COUNTERPARTS on the Bernie Bus exuded the unexpected cockiness of covering a budding insurgency, the Hillary press mimicked the morose march of our assigned campaign. Jason Horowitz, who'd also been in Hillary's 2008 traveling press when he was with the *New York*

Observer and who had his own scare trying to get to a John Edwards rally (Horowitz Taxi Hits Turkey, Both Lose, the *Observer* headline read), was now with the *Times* and assigned to the Bernie Bus. Jason knew my obsession with *Midnight Cowboy* and started calling me Rico, the character who dies in the film's final scene, his head resting on the window of the back seat of a bus. "Rico? Rico? Hey, Rico," Jason texted me, imitating Jon Voight's Joe Buck as he tried to shake Dustin Hoffman awake.

Bernie packed an auditorium in Decorah, telling the twenty-three hundred people, "Today, the inevitable candidate doesn't look quite so inevitable." Hillary, meanwhile, spoke to 450 in the city's Hotel Winneshiek ballroom, where the crowd of mostly the over-sixty-five set wore red T-shirts with the fighting words DOES YOUR CANDIDATE HAVE A PLAN FOR SOCIAL SECURITY? In Sioux City, Bernie filled the Orpheum Theatre.

Days earlier when Hillary, paranoid about comparisons to Bernie's crowd sizes, went to Sioux City, she held a "Fighting for Us" town hall at the Orpheum Theatre. Not in the theater auditorium itself, but in its ornate foyer. Supporters squeezed onto the stairs and hung over gold-leafed balconies festooned with American flags. Afterward, campaign aides bragged that a crowd stretched around the block ("at least a couple hundred people") who wanted to see Hillary but couldn't fit inside. "Shit," I thought. "If only there'd been a larger venue, like a theater, nearby . . ."

HILLARY'S IOWA TOWN halls became so frequent and intimate that they started to take on the familiar, if laborious, feel of catching up with an old girlfriend who cites GDP statistics over brunch.

On a Saturday afternoon, in Clinton in eastern Iowa (motto: "So many things to do—With a river view!"), Hillary cracked herself up when she told the small crowd at Eagle Heights Elementary, "You didn't have to name it [Clinton]. I would've come anyway!"

Her brow grew deliberate. "I got to tell you, I did a little research and Clinton County is named for DeWitt Clinton, the sixth governor of New York, and what is so interesting, because I admire DeWitt Clinton, he was the person who said, 'We're going to build a canal from the

Hudson River to Lake Erie, all the way across New York to open up the West to commerce . . .'"

It didn't matter to Hillary that by this point the crowd had started to fidget and look down at their phones. Or that most of the press, sensing a prolonged history lesson, had stood up from our row of seats at the back of the auditorium and moved to a nearby room set up with bottled water and bags of chips.

"He started when he was mayor of New York City just pushing, pushing, pushing, as hard as he could, and finally on the Fourth of July 1817 they broke ground. It took eight years. He was elected governor. He worked really hard, then he ran into some political headwinds, I know a little about that. [Some laughs.] He was voted out and then he came back. I know a little about that, too. [Some more laughs.] And then in 1825 after those eight years, the Erie Canal was opened up . . ."

Of course. Hillary *was* DeWitt Clinton. She had the perseverance and the political headwinds and the $275 billion infrastructure plan. What did Bernie have? She had so much fun telling this story that I figured we'd driven the three hours from Des Moines to Clinton that morning only so she could riff on DeWitt ("no relation"). "I think it's pretty interesting that the folks who settled here named this part of Iowa for DeWitt Clinton," she said, in conclusion. "They understood that he was a leader who set big goals and then he worked. He did the politics."

I wasn't entirely sure how DeWitt's big goals squared with Hillary's other campaign promise at the time, which was that she wouldn't overpromise. "I would rather underpromise and overdeliver," she told 460 people at the Five Flags Center in Dubuque.

The underpromise line made Brooklyn cringe. It didn't take a room full of pollsters to know that American voters preferred to elect charismatic men who wildly overpromise. But Hillary didn't want to be like them. She was a realist, or as I called it, a radical incrementalist. She'd tried to tell voters in '08 that Obama couldn't deliver on the "Hope" and "Change" he was selling. "Now, I could stand up here and say, 'Let's just get everybody together, let's get unified, the sky will open up, the light will come down, celestial choirs will be singing, and everyone will know we should do the right thing and the world will be perfect,'" she'd told a crowd in Providence, Rhode Island, during her primary fight against Obama. Eight years later, Hillary privately blamed the country's anger in no small part on Obama's inability to deliver. So for a while, before her aides pried the underpromise line off her lips, Hillary would tell Iowans, "I don't want to overpromise. We don't need any more of that."

You had to give it to Hillary for being back in Iowa at all. It had to be agonizing to get up every day and try to win over the voters who had handed her a mortifying third-place finish in 2008. For years pollsters warned her, "They just don't like you in Iowa," but had she skipped the caucuses, we all would've written stories calling her entitled, an imperial candidate, running scared from the liberal base. She wasn't going to let that happen.

Hillary tried out comedic shticks and did impersonations that I almost never saw her do outside Iowa. She'd get to the part of her stump speech about how she planned to improve the Affordable Care Act, including how she'd lower the cost of prescription drugs, a winning issue with her base of aging boomers. Part of her plan, she explained, would stop pharmaceutical companies from receiving tax credits for advertising on TV.

At this point Hillary would swirl her arms and segue into reciting, in florid detail, what sounded like a Cialis ad. "You know the ads, they have people walking through fields of wildflowers, walking on beaches, they have the name of the drug, which you know is unpronounceable, and then in a low voice . . ."—and Hillary would soften her voice, pull the microphone close to her lips and say in a deep guttural pitch that always made the Travelers look up from our laptops and chuckle—"If you take this drug, your nose will fall off . . ."

On the rope line, when a French journalist shouted over Katy Perry's "Roar" and thrust his camera Hillary's way, knocking me in the head a couple of times, "Madam Secretary, for French TV, for French TV . . ." Hillary waved and put on a faux French accent. "Itzzz zoo good to zeeeee you. Bonjour. Bonjour. French TV, bonjour."

After these town halls, Hillary stuck around to shake hands and dole out compliments. "I love that outfit!" she said, tugging at a woman's knitted scarf. "This is pretty. Is that attached? It's really pretty." She made small talk. Asked what kind of music she liked, Hillary swayed a little and shouted over a Kelly Clarkson song, "You know what, I'm kind of a sixties person to be honest. Old school, yeah, old school that brings back a lot of good memories."

Shouts of "Madam President" always made Hillary beam. "Doesn't that sound good?" she'd say. "Let's make it happen!" Hillary practically dove at a man in a gray leisure suit who carried a copy of *Hard Choices* under his arm. She signed, "Best wishes, Hillary," and as she handed it back with a come-hither wink, said, "It's a complicated world, isn't it?"

After speaking at a bowling alley in Adel, Hillary was so swarmed with a group of teachers ("I hope you get an excused absence today!") that she grabbed my phone right out of my hand to pose for a selfie. Huma whispered in Hillary's ear, "That's Amy's," and Hillary handed it back so fast it looked as if she'd suffered electric shock. "Is that yours! Oh no!" she said.

She would dispense policy prescriptions, pausing amid the crush of selfies to ask Iowans about their "COLA" (cost of living adjustments on social security) and whether they'd signed up for an "income contingent repayment plan." I once saw Hillary criticize Bernie's college plan ("I'm not going to take care of rich people") to a thirteen-year-old whom she then referred to HillaryClinton.com to read the details of her "New College Compact."

"That's what it's called, okay?" Hillary said, crouching down to eye level with the teen. He stared blankly. "Want a selfie?!" she asked.

1/29: Des Moines → Dubuque → Quad Cities
1/30: Ames → Carroll → Cedar Rapids
1/31: Council Bluffs → Sioux City → Des Moines

On late-night rides, NPR's Tamara Keith would call her three-year-old to read him *The Very Hungry Caterpillar*. Everyone thought this was adorable. But all I heard as Tamara repeated in her radio voice, "In the light of the moon, a little egg lay on a leaf . . ." was *Your fertile window is closing.* I'd turn up my music to block it out.

On long drives, our conversations started to revolve entirely around Hillary's Iowa idiosyncrasies. In '08, Hillary talked about "the nurse from Waterloo" so regularly that the traveling press had lengthy hypothetical conversations about this romanticized nurse in Waterloo, always stretching out the *loo* for several syllables the way Hillary did. Eight years later, an enormous 3-D printer had become the new nurse from *Waterluuuuu.*

Hillary discovered the printer ("the largest in North America") by accident. It was December, and she was touring Cedar Valley TechWorks in Waterloo. She watched entranced as the contraption spit out a two-foot-tall, sand-and-resin three-dimensional version of her "H" campaign logo made entirely of discarded corncobs. "Oh, come on! Come on!" The printer might as well have produced a handful of fully formed Hillary superdelegates. The 3-D printer was made in Germany, but it quickly became Hillary's favorite symbol of American exceptionalism. It was 3-D printer this and 3-D printer that five or six times a day, usually followed by her lengthy proposal to create advanced manufacturing jobs in the Midwest.

"When I went to Cedar Valley TechWorks, I saw the biggest 3-D printing machine in all of North America," she told a crowd in Waterloo. "It's amazing."

In Dubuque, Hillary called the 3-D printer "a job magnet for the Midwest." In Urbandale, she called the $1.5 million gadget "thrilling," "a big job multiplier," and "a business growth strategy."

Hillary vowed to be the president who helps Iowa "make this kind of machinery, 3-D printers in America" and, in keeping with her promise to only make modest promises, she even vowed to cut the ribbon on the first 3-D printer production plant.

After the first couple of days Hillary had relayed the story so many times that she started to mix up the details. "I was at the Black Hawk community college. They bought the biggest 3-D printer in North America because they're thinking about the future," she said in Des Moines,

Dubuque, and half a dozen other cities . . . Black Hawk College was in Moline, Illinois, and Hillary had never visited. But even (or especially?) with the muddled details, the giant 3-D printer became emblematic of Hillary's campaign style. She could be so pedantic in expressing her sincere optimism for the American worker that she either bored audiences or went over their heads entirely.

On the bus, the Travelers were simultaneously tired of hearing about the 3-D printer and at a complete loss for anything better to talk about . . .

"Hillary won't stop talking about that fucking 3-D printer."

"It'd be funny if she started placing it in whatever state she's campaigning in at the moment."

"I was just at Henderson County Community College where they had the WORLD'S LARGEST 3-D PRINTER."

"That would redeem this whole humiliating ordeal of a campaign."

25

You Will Look Happy

Thing about Iowa—no one could call it.
—RICHARD BEN CRAMER, *WHAT IT TAKES: THE WAY TO THE WHITE HOUSE*

DES MOINES → MANCHESTER, CAUCUS NIGHT, 2016

By noon on Tuesday it was still too close to call Iowa. We landed in Manchester after 3:00 a.m., dropped our luggage at the Marriott Courtyard near the airport, and had to be in Nashua six hours later for a "Fighting for Us" rally.

Bill Clinton summed up the campaign's mood after Iowa when he took the stage in Nashua and opened with this stirring line: "Well, we're here and we're awake."

We were there and we were awake.

Twenty-four hours earlier, on my last day in Des Moines, I'd spent the afternoon at the Marriott working from my Formica desk on two versions of my caucus-night story. I had a lyrical eight hundred words prewritten that assumed Hillary won. Quotes from friends gushed that the winning results showed Hillary had expelled the ghosts of 2008. She was a new-and-improved candidate with a well-oiled campaign. I assured Brooklyn the story wouldn't run until the results had been tallied, so aides told me (embargoed, on the record) that Hillary's Iowa win proved Bernie had no path and that Hillary would "handily defeat" the Republican nominee (Jeb or Rubio, obviously). "Mrs. Clinton became the first woman to win the Iowa caucuses . . ." I was writing the precursor to the November FWP story, the story all these years were leading up to.

By noon, the campaign assured me (not for attribution) that their internal polling put Hillary on track to win the caucuses by five or six points ("outside the margin of error") so I put minimal effort into pre-writing the Hillary-loses-Iowa-*again* version. I cobbled out a quick hypothetical lede:

> Des Moines, Iowa—Hillary Clinton confronted an unexpected and devastating loss in the Iowa caucuses on Monday night, thrusting her campaign into strategic upheaval and raising questions about whether her candidacy can address an angry and restless electorate.

As the caucus results trickled in, I sat there typing and deleting, typing and deleting, typing and deleting. My print deadlines came and went. In five hours, I'd maybe written fifty words of usable "B matter." It was almost midnight on the East Coast when an editor broke the news that my story would be banished to the two words a *Times* reporter never wants to hear: web only. (Despite all our talk about the web and "digital first," the six most beautiful words in the English language remained, "They want it for the front.") I should've prewritten a Hillary-says-she-won-but-basically-tied-and-we're-still-not-sure-but-let's-just-get-New Hampshire-over-with-and-move-on-to-the-primary-states-with-black-and-brown-people version, but my sources had been so certain. The polling data put her outside the margin of error.

The campaign had even less idea about what to say. Hillary arrived at her victory rally at Drake University a couple of hours late wearing red. I stood on a folding chair in the back of the room and marveled at how good Hillary, hell, the whole family, was at smothering any honest emotional reaction. With Bill and Chelsea standing onstage behind her, Hillary drew into some deep reserve of fakery and willed herself into looking happy, as if trying hard enough would make it so. I remembered the advice Huma had given to a sobbing Anthony Weiner staffer after the Carlos Danger sexting scandal enveloped her husband's mayoral bid. "I assume the photographers are still outside, so you will look happy?" Huma said.

With a plastered-on grin, Hillary pointed at the crowd, equal parts Iowans, Washington insiders, and New York donors and a pile of baby boomers who came up in caravans from Arkansas. I saw Hillary smooth her suit jacket. She ran her palms down the sides of her thighs, the kind of fidgety gesture she hardly ever made onstage.

"I love it! Wow, what a night, an unbelievable night," she said, letting the word "unbelievable" hang there. *You will look happy.* "I stand here tonight breathing a big sigh of relief. Thank you, Iowa!"

Rachel Platten's "Fight Song" came on, ending Hillary's six-minute-and-forty-five-second speech. "LOADING!" a rosy-cheeked campaign staffer yelled over the campaign's newest girl-power pop anthem. We rolled up our power cords and trudged through the parking lot to the bus with our open laptops cradled in our arms. The Travelers looked aghast.

"Um, what just happened?"

"That didn't feel like a victory rally."

"No, no it did not," I said.

We were sitting on the bus in the parking lot when my phone exploded with texts and emails. Democrats saw the virtual tie as an omen and wanted campaign manager Robby Mook "layered." Nobody said he should've been fired; that would've led to too many negative headlines. Nothing drove news traffic like Clinton infighting.

The Democrats who'd come up during the McGovern campaign, in particular, worried the thirty-six-year-old campaign manager's approach—all math and no poetry—needed to be replaced by some old-fashioned fire. (The younger operatives would point out that Nixon defeated McGovern in a landslide.) People proposed Maggie Williams, who had been Hillary's chief of staff in the White House and was one of the only people who could tell Hillary no. Maggie had been dragged into easing Clinton melodrama for years, including reluctantly taking over in 2008 from then-campaign-manager Patti Solis Doyle who was fired after Iowa. To avoid negative headlines, Robby could even keep his title and his corner office and his standing desk and his "mafia" of obedient bros who'd followed him from McAuliffe's Virginia governor's race. But Maggie had been there, done that. She was content to offer outside counsel from her perch at Harvard.

By 2016, I'd forgotten most of the katakana alphabet that I'd attempted to learn in Japan, but I did speak fluent Political Cliché. I turned again and again to the most overused words of the Democratic primary: *Organization* and *Enthusiasm*. I asked Bernie and Hillary people which one was more important. They all said, "You need both." But like a bratty teenager playing a drinking game, I demanded an answer: "You MUST choose." Most people settled (not for attribution) on Enthusiasm. "Organization don't mean shit if people aren't excited about the candidate," a veteran Texas Democrat said. Robby was an Organization man; Bill Clinton, the ultimate Enthusiasm guy. "What's the data and organization for if voters don't like Hillary?" Bill would say to anyone who would listen. "They need to see the person I know."

About a week before the caucuses, at the end of an epically newsless bus swing, Demi Lovato performed "Confident" on campus at the University of Iowa in Iowa City, the epicenter of the "Feel the Bern" movement. She introduced Hillary, saying there wasn't "a woman more confident than Hillary Clinton." (Telling the crowd the truth—that one of Hillary's more endearing qualities is that despite her successes she is a heaping pile of insecurities—wouldn't have played well.) Hillary came onstage to a sea of Snapchatting coeds. She thanked Demi and spent a total of 3.5 minutes reminding the couple thousand students in the audience to caucus.

I waded into the crowd afterward. I didn't meet a single student who said they were supporting Hillary. "I'm just here for Demi," a rail-thin sophomore named Tyler told me. So why was he wearing an H sticker? He looked down at his plaid flannel shirt as if it surprised him to see it stuck there against his almost inverted chest. "I don't know. They gave it to me. But Demi's cool, she's got my vote." We heard the same responses at Katy Perry concerts and Lena Dunham house parties.

"I'm here for Lena," said Heather, a thirty-three-year-old from Cedar

Rapids. "I don't want to vote for someone for president of the United States because I love Lena Dunham."

Pete D'Alessandro, Bernie's top man in Iowa, could hardly drum up an endorsement from anyone other than Susan Sarandon and Mark Ruffalo. But when I asked him about Enthusiasm vs. Organization, he compared these star-studded Hillary events to a story from the 1968 campaign. Pete had slithery hair and a goatee and none of the polish of the political class who worked for Hillary, but he spoke about the race with a Zen certitude that I never heard from the Clinton camp, a Jedi master in Dickies and shabby black fleece. In '68, Eugene McCarthy's campaign held a barbecue that attracted hundreds of voters lined up with MCCARTHY FOR PRESIDENT signs as far as the eye could see. The scene led Bobby Kennedy's campaign to think McCarthy had a lock on the primary. Kennedy ended up defeating him badly.

"Turned out," Pete said, "they just came for the ribs."

I EVENTUALLY SETTLED on a nut graph that said the uncertain outcome had "dealt a jolting psychological blow to the Clinton campaign, leaving volunteers, donors and aides confused throughout the night, and then crestfallen."

Our newspapers and networks paid $1,700 per person for us to take the three-hour press charter flight from Des Moines to Manchester. Hired Gun Guy strong-armed our bosses to call Iowa in Hillary's favor while swearing to us that she'd *obviously* won. I didn't think I could experience a middle-of-the-night charter more fraught than the flight to New Hampshire after Hillary's third-place finish in 2008, when the campaign, after blowing $50 million and nearly half a year there, collectively pooh-poohed the state. "The worst thing would be to overcount Iowa and its importance," Mark Penn told us on the let's-get-the-fuck-out-of-Iowa caucus-night charter to New Hampshire eight years earlier. "Iowa is so small. It's like a mayor's race in a medium-size city," added one of the '08 Guys. At least back then the campaign's spin was quotable. The shit sandwich the '16 team tried to feed us was virtually useless. No matter how hard he pushed, and how slick his delivery, something about Hillary's Hired Gun Guy telling us "We believe strongly that we won tonight" didn't exude triumph.

The press climbed over the seats and poured into the aisle. We steadied ourselves on each other's shoulders and hoisted each other up until our muddy snow boots stood on the leather business-class seats. As photographers and cameramen affixed their lenses above our heads, the print reporters squatted down on the ground, our slingshots loaded. "Can you move your hair?" a photographer yelled at me in a request that would become almost as frequent as "What's the Wi-Fi password?" We were ready for our $1,700 gaggle. It went something like this:

[truncated audio]

HIRED GUN GUY: We've heard a lot in the last few weeks about Enthusiasm, but it turns out there's a heck of a lot of Enthusiasm for a progressive who will get things done.

TRAVELER #1: Is your internal polling off? It seems like this came as a surprise. It had you up . . .

HIRED GUN: What do you mean?

TRAVELER #1: A tie isn't what you were expecting over the past few days . . .

JEN: We believe we won.

TRAVELER #2: Why didn't you declare victory in that speech?

HIRED GUN & JEN: We did.

TRAVELER #3: You guys had a ground operation [i.e., Organization] that was second to none. What does it say about the candidate and her message? Was there a problem there?

HIRED GUN: We believe she won.

And so on for nine minutes and thirty-one seconds until Jen announced, "The rest of the plane will be off the record."

I heard from my editors at the *Times* who were stuck in the Waterloo conference room because of a looming blizzard. They needed an official campaign comment for my story before we took off. I made the very bad decision to yell at The Guys, "Are you bringing in any new advisers to the campaign? Layering over Robby?" I asked this not once, not twice, but three times. "I said we're now off the record," Jen shot back as she walked toward the staff seats at the front of the plane. Then Brown Loafers, who of the current crop of Guys took things the most personally, unleashed all his pent-up rage about the *Times*, and my coverage, into one drop-dead look and three irascible words: "Are you serious?"

I was always misreading the ebbs and flows of the press scrum. After the *Guernica* press conference, I almost never asked about Hillary's emails, figuring the TV reporters who were jumping out of their chairs with "WHAT ABOUT YOUR EMAILS?" questions had that covered. I'd scream softballs ("What kind of ice cream did you order?") when my editors needed answers about some shady Clinton Foundation donor. Or I'd shout a question about foreign policy ("Do the reports of ISIS holding Yazidi women as sex slaves change your thoughts on intervention?") over the mooing of dairy cows at the Iowa State Fair.

In 2008, Hillary used to tell me I was asking the wrong questions. "Well, Amy, that's not what you should be asking . . ." and then proceed to respond to the question she wanted to answer. "God, I wish you weren't always asking the wrong questions," Anne Kornblut, the *Washington Post* reporter, would say as the scrum disassembled.

I botched even the most straightforward questions. In South Dakota in 2008, I raised my hand to ask Obama a question about his faith outreach adviser, Joshua DuBois, and whether he could win over evangelicals. But I'd accidentally said Jeremiah Wright.

"Senator, Jeremiah Wright told me—"

"Jeremiah Wright," Obama said, taken aback.

"I mean Joshua—"

"You mean Joshua DuBois?"

At this point the entire press corps was laughing so hard that I don't think Obama heard me when I muttered, "Right."

Late that night, on the "Change We Can Believe In" plane, Obama came back to the press quarters. I hadn't met him yet and introduced myself. "I know you," he said, with a giggle. "You're the one who asked me about Jeremiah Wright."

BUT ASKING ABOUT Robby's potential ouster in front of the entire traveling press corps before the caucuses had been officially called, or any alcohol served, had possibly been worse. Hacked emails later showed that Brown Loafers had called me an "idiot" to one of Hillary's closest friends, Capricia Marshall, after she'd asked about my reporting on Robby. I would've thought he'd called me worse. (I did get a nice note from the Trump campaign saying that Mr. Trump didn't think I was an idiot.)

We must've been somewhere over Ohio. A couple of photographers and sound guys were on their third lukewarm can of Coors Light. I chased Brown Loafers down the aisle to try to explain myself and hopefully smooth things over.

"Would you rather I don't ask you about rumors I'm hearing?"

He kept walking.

I tried again. "We should really talk about this . . ."

He didn't turn his head. He walked straight to the back of the plane and plopped down with the Tripods, whispering to them over red wine in plastic cups and reminding me, without saying a word, of my own minusculitude. That flight would draw a line in the sand between me and The Guys that a couple of days later in New Hampshire would morph into an insurmountable moat.

26

He Deprived Her
of a Compliment

MANCHESTER, FEBRUARY 2016

"'Psychological blow'? Are you in my head, Amy? No, really, tell me, are you in my fucking head?" His eyes protruded, a web of veins pulsing from his forehead. I knew it had to be bad.

Of all The Guys, this was Hillary's measured, policy-minded one. I never knew this Guy to confront reporters over much of anything, except maybe to point out Hillary's role in the Iran nuclear deal. ("When Iran was serious about coming to the table, we had laid the table. By this point it was Secretary Kerry's turn . . .") But in the lobby of the Manchester Radisson three days after the Iowa caucuses, he saw me, stopped his deliberate stride, turned around, and unleashed.

By 1:00 p.m. the day after the caucuses, during the "We're Here and We're Awake" rally in Nashua, the AP had called Iowa for Hillary. She'd defeated Bernie by one-fourth of one percentage point, including, we later learned, winning six coin tosses in tied precincts. The future of our democracy dangling in the air as Iowa's state coin descended in the Weeks Middle School gymnasium, home of the Wildcats.

Tails. Hillary gets the extra delegate.

Hillary couldn't stand when reporters "put her on the couch," a common practice given the "opaque reality" of her own self. Even Hillary's closest girlfriends admitted she was impenetrable, so what the hell did some reporter who was in high school during the Clinton administration know?

Aside from the "psychological blow" stuff, I heard that Hillary was livid about the "tone" of my Iowa story. She went on a tear to The Guys that only she, Hillary Clinton (she'd recently dropped the Rodham), sufferer of double standards, endurer of an impossibly high bar, number one most put-upon politician in modern history, could win Iowa and still have the *Times* say the caucuses "dealt a jolting psychological blow" to her campaign.

I couldn't argue with that. But she was also Hillary Fucking Clinton and she essentially tied with a seventy-four-year-old Brooklyn-born Jew whose primary legislative achievements in a quarter century in Congress were the renaming of two post offices in Vermont. Now we were in New Hampshire where polls had Bernie up by as much as thirty points. Brooklyn stopped spending money in the state and aides urged Hillary to skip it and focus on Nevada and South Carolina. But New Hampshire had done a lot for the Clintons, and she would need its four Electoral College votes in the general.

Hillary became sullen. There is an image of her with her head on Bill's shoulder. He has his reading glasses on, and she is limp and resting on him like she doesn't want to do it anymore. She was pouty and aggrieved but not surprised that the media hadn't given her rightful due. Brooklyn summed up her sentiment as "I'm the first woman to ever win the Iowa caucus and it's like it never happened."

I don't regret the story or its tone. At the time, the caucuses looked like a tie and the campaign contemplated an upheaval. But I regretted the confrontation. I regretted that a couple of days into New Hampshire things had become so toxic with The Guys that they hardly spoke to me except for the most mature of the bunch, Policy Guy—the one who was also personally closest to Hillary—cornering me in a hotel lobby to spontaneously combust.

I tried not to meet Policy Guy's eyes. To my left there was a wall of windows, and I envisioned making my escape, busting through leaving an outline of my body in the glass like a cartoon cutout. I stammered, "Uh, well, um, I'd be uh happy to discuss your concerns with the story, or uh if you want to talk to my editor . . ."

THE LOBBY SMELLED of burnt breakfast sausages from the buffet that a couple of busboys were now packing up. The invite to the "Bloomberg Politics Breakfast Briefing with Robby Mook" read, "The event is a seated buffet breakfast, so **please arrive by 7:15** to ensure that we can start on time." I got there at 7:45 a.m. dreading seeing Robby, whom I'd just eviscerated to several million unique viewers.

As usual, I tried to comfort myself with the free food, piling two halves of eggs Benedict topped with sauce as bright and yellow as antifreeze onto my plate. I took an empty seat between Margaret Talev of Bloomberg and MSNBC's Kasie Hunt at the end of the rectangular table ensconced in a black tablecloth and a white silk runner down the middle. The *Boston Globe*'s Annie Linskey, the *Washington Post*'s Ruth Marcus, and *USA Today*'s Susan Page sat nearby. The men, Bloomberg's Al Hunt, John Heilemann, CNBC's John Harwood, and others were clustered together in the middle of the table as Robby and Hired Gun Guy held court, giving the room a gender divide as pronounced as at my third cousin's Chabad-Lubavitch wedding ceremony.

"My question is about the younger voters who seem to be in full-on rebellion against Hillary Clinton. I've never seen a seventy-point margin in a demographic in a primary like this—" one reporter started to ask.

Robby interrupted, "Well, first of all, that's based on an Iowa entrance poll, and . . ." (Days later, Bernie would win the New Hampshire primary by a sixty-nine-point margin among voters under thirty—this wasn't an Iowa entrance-poll problem.)

Robby, with his boyish good looks and buoyant, affable manner, had become somewhat of a sex symbol among a very specific subset of Washington power gays. He didn't flaunt his position as the first gay campaign manager of a major presidential candidate. But Robby had a remarkable ability to move his lips without saying anything. We'd ask him questions, and words like "she's going to fight for every single vote" and "Is the next president going to keep us pushing forward in the future or are we going to go back?" would come out. A couple of prominent gay political reporters started to refer to Robby as a cute robot "assembled in a closet in Vermont."

Asked, again, why young voters weren't supporting Hillary, Robby said, "You'd have to ask the voters that." Hired Gun Guy—sensing a malfunction—earned his keep. He staged a hostile takeover, picking up the toughest questions and answering with his usual smooth élan of a spokesman for BP after the oil spill.

REPORTER: Do you or do you not think there's an element of sexism in the way he's [Bernie] talking about Hillary Clinton?
MOOK: I'm going to leave—I'm going to leave you guys to judge that.
HIRED GUN GUY: Can I make one comment on that?

Then Hired Gun, who looked less and less like Ben Affleck as the campaign wore on, with dark circles under his eyes and the pasty skin of someone who spent too much time in a drab office building in Brooklyn and a Manchester Radisson, went into a lengthy presentation about Bernie's "troubling adoption" of the same attacks lobbed by the so-called Bernie Bros.

"It can be vitriolic," he said. "And I think the Sanders campaign needs to be aware the extent to which, in an effort to mobilize and galvanize their supporters, they let the mentality or crudeness [of the Bros] seep into their own words and criticisms that they hurl at Secretary Clinton."

After Iowa, the Bros, a vile caboodle of loosely organized online trolls who harassed anyone (particularly those of us with vaginas) who seemed disloyal to Bernie, became emboldened. Every time I wrote about Hillary, I heard from enraged Bros furious that the media had discounted their man. They flooded the *Times* comments section and wrote letters to the editors littered with four-letter words. They regularly addressed me as a "NY Times Presstitute," a "donkey-faced whore," and a "life-support system for a cunt" who works for the "JEW York Times."

People started to whisper that Brooklyn had invented the Bros in some dark-arts effort to drive young women to their rightful place, With Her. I like a good Clinton conspiracy as much as the next reporter, but Brooklyn didn't create the Bros. It took six staffers and a focus group to compose an official tweet, signed with an "H." ("FOR APPROVAL: Slight Edit to

Tweet," the emails read.) There was no way Hillary's campaign had the creativity to come up with "NY Times Presstitute."

Now Hired Gun Guy was trying to tell us that Bernie and his Bros were one and the same because Bernie had the anti-woman audacity to claim Hillary wasn't a true progressive. The offending line came at a CNN forum the day before the breakfast when Bernie said, "Some of my best friends are moderates. But you can't be a progressive and a moderate at the same time."

As Hired Gun Guy made his case, the press rolled our collective eyes so far back in our heads that the first question that popped in my mind was from the movie *Heathers*—"Did you have a brain tumor for breakfast?"

Somebody let out a groan and said, "C'mon. Do you think the umbrage really fits the comment here? I mean, he was asked if he thought she was a progressive. He basically said, 'Well, she said she was a moderate and she says she's a progressive and you can't be both.' That's not exactly the insult of the century."

"I would just put it back and say what's the intent in making the comment? . . . In trying to deprive her of the compliment it is to be called a progressive?" Hired Gun Guy said. Robot Robby sat silently, his lips pursed into a heart-shape.

I hadn't realized Hillary was such a delicate flower, but I understood the frustration. She'd spent much of her career trying to overcome a caricature of the liberal feminist. She'd written her senior thesis on Saul Alinsky and told the Wellesley women of '69 that politics was "the art of making what appears impossible, possible." As first lady, she'd been the leftist foil to her husband's centrist agenda. She'd privately opposed NAFTA, pushed for universal health care, lamented the welfare overhaul that gutted federal assistance of the poor by $55 billion and effectively ended her friendship with her mentor Marian Wright Edelman. The West Wing called Hillaryland "the Bolsheviks."

But—even in a primary race where she could have used her progressive past to help brush back Bernie's attacks from the left—Hillary never mentioned the Bolsheviks or any of this, too afraid of sounding like she opposed her husband's administration and of rekindling the old commie caricature that she'd worked so hard to shake and that would haunt her in a general election. *We are all forged in the crucible of our mistakes.*

I had to practically pound on the table to get Robby and Hired Gun Guy to take a question from the estrogen-heavy end of the room. I asked if Hillary had expressed to her top aides her frustration that she spent most of her career trying not to be a liberal cartoon and now she was trying to convince voters she was a "progressive who likes to get things done." The way I remembered it, I phrased this question brilliantly, inserting cultural nuances and gender dynamics into an otherwise dry political debate. But when I looked back at the transcript, this is what I actually said: "Has she ever, like, expressed to her aides frustration that, like Jesus, the whole early part of my career was [inaudible] that I wasn't like a scary [bra-burning liberal] [inaudible] we need to convince them that I have been there on all these issues?"

Snorts from the middle of the table. *Heh. Heh. Heh.*

"Yeah, when is she going to burn her bra?" Heilemann asked

Heh. Heh. Heh.

The breakfast had descended into an episode of *Beavis and Butt-Head.*

"Are there burnt bras in the archive in Little Rock? If so, when will we see them?"

Heh. Heh. Heh.

All Robby could say was "Wow."

AFTER THE BRA Breakfast, I wanted to run away from the Radisson, get in my Subaru, and drive to the barre class in Bedford that I'd found the day before. A former state representative taught the class in a carpeted attic converted into a workout studio. Christmas lights hung overhead, and clients affixed paper hearts for Valentine's Day to the walls, with messages like I LOVE ME and I ♥ GOD.

Instead, I ended up in a lobby showdown with Policy Guy over a web only story that had long been lost in the news cycle and the deluge of post-Iowa, pre–New Hampshire hot takes.

I needed to defuse the situation, to stop being so defensive, and just give him the apology he wanted to hear. But when I did, the words came out as sincere as Hillary's "I'm sorry that this has been confusing to people."

"Look, I hear you," I said. "I'm sorry you hated that story, but my meter ran out, and I've really gotta go—"

"No, tell me," he said, swerving his neck forward and down to meet my eyes as I pretended to dig around for my keys. "Are you in my head?"

Apparently so.

I WOULD'VE HAD more patience for this ass ripping had the Clintons not simultaneously sponged off my reporting. I spent months reporting a feature about how Hillary had gone undercover to investigate school segregation in the South.

> DOTHAN, Ala.—On a humid summer day in 1972, Hillary Rodham walked into this town's new private academy, a couple of cinder-block classrooms erected hurriedly amid fields of farmland, and pretended to be someone else.

The Guys, as usual, doubted my motives and tried to kill the story. It ran right around Christmas on the front page, the primest of prime real estate the Clintons, a couple of *Times* print subscribers in their late sixties, could've hoped for. By February, at the same time Brooklyn was ready to string me up in a plaza in Manchester, the Clintons had turned my reporting into full-blown agitprop.

Hillary was twenty-four when Marian Wright Edelman's advocacy group was investigating the hundreds of private schools that sprung up across the South to cater to white families after the 1969 Supreme Court decision forced public schools to integrate. She went to Dothan, a town near the Chattahoochee River and the Fort Rucker Army Base named after Genesis 37:17: "They have gone away, for I heard them say, 'Let us go to Dothan.'" She pretended to be a housewife who wanted to enroll her son in the town's new private academy.

"I went through my role-playing, asking questions about the curriculum and makeup of the student body," Hillary wrote in *Living History*. "I was assured that no black students would be enrolled."

For whatever reason, Hillary didn't like to talk about Dothan. She devoted only three hundred words in her 567-page memoir to the experience. Even during the 2008 primary amid accusations of racism, Hillary didn't talk about her undercover work. ("Ya think that would've been good to know?" a Guy from the '08 campaign said after my story ran.)

When I initially told The Guys I wanted to do a deeper piece about this brief chapter of Hillary's career, they assumed I'd make her look shady or underhanded. "You always find a way," they'd said. She'd recently told voters at a diner in Manchester that the Marine Corps turned her away in 1975—biographical color that led to at least three days of mockery and "Two Pinocchios" from the *Washington Post*'s fact-checkers.

I cried, actually broke down in my mother's bathtub in San Antonio over Thanksgiving weekend, after a Brooklyn source told me Hillary hadn't even considered my months of interview requests.

"But, but, this is a positive story. I'm not trying to screw anyone," I said between sobs. "And, and, she just gave an interview to the *Boston Globe*. Why not the *Times*?"

"Oh, yeah," he said. "Well, Annie [Linskey] went to Wellesley so she wanted to do it."

Now, with Hillary about to lose New Hampshire and reliant on a firewall of black voters in the South, my "Undercover Hillary" story became political gold, tangible evidence on the front of the *Times* that she'd been at the forefront of civil rights. She *was* a progressive who liked to get things done.

Without mentioning me or the *Times*, Bill started to refer crowds to "a very nice article in the press." He painted a portrait of Hillary as a young activist, unafraid and on a mission in the Deep South when women didn't do such things. He added a few factual flourishes.

"She posed as a housewife who had just moved to this little town in Alabama, and you know, she made small talk with the guy at the school . . ." I heard Bill say at an event in Rochester, New Hampshire. "It took guts forty years ago to do that and it changed and they lost their tax exemptions and they had to change all their practices."

(Houston Academy, the private school Hillary visited—and that I toured in the course of my reporting—didn't lose its tax-exempt status. In 2013, eight of its 527 students were black.)

Bill loved the "Undercover Hillary" story so much that months later he made it a central part of his prime-time address to the Democratic National Convention in Philadelphia.

"They exchanged pleasantries, and finally she said, 'Look let's just get to the bottom line here, if I enroll my son in this school will he be in a segregated school, yes or no?' And the guy said absolutely," he told a hushed Wells Fargo Center. "She had him!"

I may have been in open warfare with the Hillary who won Iowa with the help of a coin toss and who thought I was an insipid bottom-feeder, but I adored the Hillary who went undercover that summer in Alabama. The Hillary who met with civil rights leaders in Atlanta and rented a car and drove to Dothan where she checked herself in at a Holiday Inn off Ross Clark Circle.

I pulled up my fur-lined hood and made my escape back into the grimy Manchester air. I heard the salt and ice on the sidewalks crush beneath my boots. I wanted to believe that this Hillary would've kind of liked me, too.

27

"Saint Hillary"

FLINT, MICHIGAN, FEBRUARY 2016
"Saint or Sinner, Moralist or Machiavelli, Mother Theresa or Lady Macbeth," Diane Blair wrote in a 1996 journal entry. "Hillary, like most of
us, some of both and much in between."

The black block letters on the low-hung billboard hovering over the
awning of a MetroPCS store, the only vibrant retail I saw in Flint other
than a pawn shop and a funeral home, if a funeral home can count as
retail, read REJOICE IN HOPE. BE PATIENT IN TRIBULATION. ROMANS 12:12.

It was Sunday morning, forty-eight hours before the polls opened in
New Hampshire and 730 miles away. Most of the Travelers stayed behind in Manchester, but I'd made my way to Flint so I could go to church
with Hillary. The water crisis had become a national health emergency,
and Hillary was still one of the only politicians, and the only candidate,
really talking about it.

I waited at the end of the second to last row of purple upholstered
pews at the House of Prayer Missionary Baptist Church. The lead came
from the Flint River and the corroded pipes had pumped the brown
sludge into public schools and kitchens and bathrooms in homes that
had once been aspirational middle-class bungalows but that now dotted
the landscape of postindustrial blight. On the drive to Flint, I'd passed
a truck propped up on cinder blocks and spray painted with the silvery
sprawl of MAKE AMERICA GREAT AGAIN. If there were a travel poster advertising like a cruise getaway to this previously Great America, it would've
looked like Flint, when it was white and before the factories closed.

In the church bathroom, as Rev. Kenneth L. Stewart delivered a sermon about Flint being on "God's waiting list," I saw a little girl in braids and a white dress perch on her patent-leather tippy toes to wash her hands. I ran into the stall and sobbed. The Flint trip slapped me upside the head and woke me up from my fights with The Guys, my wrestling over bylines, all my unimportant coastal concerns.

I went to Flint mostly because I wanted to see Hillary in a different setting after so many town halls in Iowa and New Hampshire, where almost everyone was white. The Flint trip signified the next stage of the race—when she would rely on black voters. Critics would describe the visit as the most jarring example of Brooklyn's overreliance on "identity politics" and Hillary's overreliance on the black voters who elected Obama twice. But when I got to Flint, it didn't feel like pandering or identity politics. It didn't feel like politics at all.

A young woman told me she'd miscarried twins. A mother of four said her eight-year-old son had been bright, smarter than most in his first-grade class, until the lead got into his blood. A man pulled up the sleeve of his maroon suit jacket to show me the chalky white rash that ran from his wrists up his arms and wouldn't wash away. I asked Bobby Blake, a pastor at another local church, if people thought Hillary was there because she needed black voters. "I don't care why she came," Blake told me. "This town has been living out of a bottle. My question is, where's everybody else?"

FOR OVER A year, I would go to at least one black church with Hillary almost every Sunday. We went to black churches in North Carolina, Mississippi, Tennessee, Kansas City, Texas, and Brooklyn. I started to learn the rhythms. How Hillary would always step to the pulpit, take a deep, freeing Sunday morning inhale, and then open her remarks with Psalm 118:24: "This is the day the Lord hath made; let us rejoice and be glad in it." She'd talk about working to end "systemic racism" and, well before the general election, she shat all over Trump's slogan. "America has never stopped being great," she told the Greater Imani Cathedral of Faith in Memphis. When the choir belted out "Glory, Glory, Hallelujah," I couldn't help but shake my shoulders. I pretended to look down at my phone as I bowed my head and mouthed the words "Lift every voice and sing."

Between the hours of 10:00 a.m. and noon, Hillary became a different person. She recited scripture with the fluency of a renowned theologian, the verve of a TV evangelist. She loved the Epistle of James: "Scripture tells us that faith without works is dead." She said she tried to live up to the prophet Micah's teachings, "that we do justice, love kindness and walk humbly with our God." She explained that her Christianity is "a journey that never ends." She described her calling as answering "the charge given to us by Jesus, as Matthew records" and heaping love on "the least, the last, and the lost."

In another life, she could've been a Methodist pastor or Princeton theology professor. She could've been a spiritual guide steering her young congregants not "to slumber while the world changed around us," as she often said she learned in 1962 when her own youth minister, Don Jones, at the First United Methodist Church in Park Ridge, took her to hear Martin Luther King Jr. speak.

On the campaign plane in 2008, I saw Hillary, late into the night, reading the leather-bound Bible that she travels with. But only once a week would she step to the pulpit and publicly draw into this divine reserve. *This is the day the Lord hath made . . .*

Hillary used to talk about her faith all the time. In a 1993 speech at my alma mater in Austin, Hillary said the country was suffering from a "crisis of meaning."

"We need a new politics of meaning. We need a new ethos of individual responsibility and caring. We need a new definition of civil society . . . as to how we can have a society that fills us up again and makes us feel that we are part of something bigger than ourselves," she said.

The Austin speech led to a *New York Times Magazine* cover story by Michael Kelly, who by then was on his way to becoming a neocon. SAINT HILLARY: MORE PREACHER THAN POLITICIAN, THE FIRST LADY SEEKS A NEW REFORMATION, CONCERNED LESS WITH HOW GOVERNMENT SHOULD BEHAVE THAN WITH HOW PEOPLE SHOULD, the headline read.

The *Times* story made Saint Hillary seem like a kooky, New Agey do-gooder. She quoted Dietrich Bonhoeffer and said that "the very core of what I believe is this concept of individual worth, which I think flows from all of us being creatures of God and being imbued with a spirit."

If there's one thing the political press can't stand, it's sanctimony. The mockery came from all sides. The *Atlantic* said the first lady had pulled on a "quasi-mystical-socio-politico-psychological coat of crazy colors."

After that, Hillary rarely talked about her spiritual side. She even hesitated to wear all white, as she had for the *Times* cover photo, until twenty-three years later when she won the Democratic Party's nomination. "Still endless press about Saint Hillary piece by Michael Kelly," Diane Blair wrote at the time. *Another thing to blame the* Times *for.*

The GOP candidates by contrast were tripping all over themselves to prove their Christian bona fides.

A few days after Trump referred to Second Corinthians as "Two Corinthians," making the eighth book of the New Testament sound like a buddy sitcom set in Athens, Maggie Haberman and I sat a few rows behind and to the left of Trump at the First Presbyterian Church in Muscatine, Iowa. Trump, his communications director, Hope Hicks, and campaign manager Corey Lewandowski arrived late to the little redbrick chapel up the street from a frozen river scattered with ice fishermen. Trump stayed for the whole sermon about the world's largest ball of twine in Darwin, Minnesota, a metaphor for unexpected gifts. ("We don't have much of a town left but that twine ball really draws them in . . .") He then dropped two crisp fifty-dollar bills into the collection plate.

When all the politics and caution were stripped away, Hillary was at her core a Methodist, a church lady, a fire-and-brimstone Jesus-saves believer. When I try to tell people this, they always say she's just another pandering politician. Trust me, you don't drop the prophet Micah in mid-conversation because you're pandering. You can't fake extended allusions to the Eight Beatitudes of Jesus (as she did during a town hall in Knoxville, Iowa) or casually quote the Jesuit academic Henri Nouwen's parable on the prodigal son during a CNN town hall (as Hillary did in response to a rabbi's question in Manchester).

"Regardless of how hard the days are, how difficult the decisions are, be grateful, be grateful for being a human being, being part of the universe," Hillary told the rabbi. "Be grateful for your limitations."

BY THE TIME Hillary arrived in Flint, her limitations were all anyone was talking about (myself included). In the finale of his sermon, Stewart declared to the congregation, now on their feet and filled with the temporary ecstasy of imminent salvation, "Hillary Clinton is on the waiting list," and "Have I got a witness?"

And the crowd shouted, "Amen!"

Hillary took her place behind a wooden podium that dwarfed her small frame so that only her head and an inch or two of her tweed jacket showed. Behind her was a choir in purple and gold sleeves and a mural of a black, bare-chested Jesus being baptized in the River Jordan. Ushers in white uniforms settled into whatever empty seats they could find. By then some local reporters and the network embeds, who had also made the trip from Manchester, filled out the back pews.

"I am here because for nearly two years, Flint's water was poisoned," Hillary said, sounding as I imagined the young activist who went undercover in Alabama in 1972 had sounded. "I'm here because for nearly two years, mothers and fathers were voicing concerns about the water's color, about the smell, about the rashes it gave to those who were bathing in it.

"Clean water is not optional, my friends, it's not a luxury," she said. "This is not merely unacceptable or wrong, though it is both. What happened in Flint is immoral."

Immoral. It wasn't a word presidential candidates, especially Democrats, usually used, but I wasn't watching a politician. This was Saint Hillary. The Sunday school teacher awakened by the left-leaning youth minister who believed in the Methodist tenet that "the ultimate test of a moral society is the kind of world that it leaves to its children." The first lady who defended her husband's infidelity in Biblical terms: "You know, in Christian theology, there are sins of weakness and sins of malice, and this was a sin of weakness."

By the time I left the House of Prayer to get a flight back to Boston and cab it back to Manchester, Flint wasn't the story. Not even close.

Bill Clinton had gone off message. He told a crowd in Milford, New Hampshire, that Bernie was a "hermetically sealed" hypocrite. "When you're making a revolution, you can't be too careful with the facts," he said.

Meanwhile, Madeleine Albright and Gloria Steinem had declared war on young women for supporting Bernie—making Hillary's seem like the embodiment of antiquated mean-girl feminism and driving even further away the very voters she'd hoped to inspire.

"There's a special place in hell for women who don't help each other!" Albright, then seventy-eight, had said at a rally, repeating a line she often used.

And Steinem, at a very sexy eighty-one, had pronounced on Bill Maher, "When you're young, you're thinking: 'Where are the boys? The boys are with Bernie.'"

Cue outrage.

My flight was delayed. The embeds and I settled into a booth at a TGI Fridays at the Detroit airport. By the time I finally got back to Manchester, it was after 3:00 a.m. I fell into my bed with all my clothes on and the dry heat of a midrange hotel cranked up on high, wondering if Flint, and Saint Hillary, had even really happened.

—
28
—

I Hate Everyone

MANCHESTER → NEW YORK CITY → LAS VEGAS, FEBRUARY 2016
The morning of the New Hampshire primary, I was bound for another web only story about why Hillary lost. I was killing myself on the road and still falling behind. My editors weren't thinking about me to write the main stories. My colleagues bulldozed me on primary and debate nights.

"Oh, it's just easier for [so-and-so] to write it because he's here," was usually how editors would explain it to me when I whined—and Jesus did I whine. I whined so much that even on a team of master whiners an exasperated editor told me as he loaded a Flavia pack into the hotel coffee maker that I was the worst. "Please, you mean I'm worse than [so-and-so] who everyone could hear screaming at you through closed doors?" Yes. "Worse than [so-and-so] who refused to leave New York for more than two days and then complained the whole time he was here?" Yes. "Worse than [so-and-so] who is usually on cable TV when everyone else is slaving away on deadline?" Yes.

I adopted Hillary's mood. I went around despondent and aggrieved, pissed off at the world, at my editors, at myself for not being "likable enough." I'd even caught the sinus infection Hillary had been fighting off. When friends would check in on me, I'd send them all a meme of Toby Ziegler, the fictional White House communications director on *The West Wing*, looking suicidal in a drab brown suit with the quote, "There is literally no one in the whole world I don't hate right now."

Jen Palmieri told me Hillary had been "working through a lot in her head" when she got introspective at a town hall at New England College right before the primary. "I am who I am," Hillary said. "I can't do some kind of personality transformation."

Same.

After nearly twelve weeks straight on the road, one caucus, one primary, and countless days eating multiple meals at Panera Bread, I'd hoped to spend some time in New York. I missed Bobby and the Lower East Side and kale. I'd been wearing the same three Old Navy turtlenecks, in black, blue, and gray, for months. But Carolyn had other plans.

In her Carolyn way, she explained, as we sat on lawn chairs in the stuffy poolroom at the Marriott Courtyard the morning of the New Hampshire primary, that she needed me to be the "Nevada bureau chief." There was no such thing as the Nevada bureau chief. She needed me to keep an eye on the Democrats while most of the team focused on the GOP primary in South Carolina. I didn't care. I was just relieved to feel the warmth of Carolyn's sunshine on me again.

I said yes and then started to explain that I needed to go home for a couple of days. "I've got to do laundry, repack, I haven't seen Bobby in weeks—" I started. I hadn't intended it, but there's a better than average chance that this sounded like whining. Before I could finish the sentence, Carolyn turned her head toward the windows and squinted. At least a foot of fresh snow had fallen on top of the grimy older snow, bringing with it the chill of Carolyn's disappointment. The air smelled of chlorine, and steam rose from a hexagon-shaped hot tub. "Well," she said, "I guess I could get Adam to do it . . ."

She meant Adam Nagourney, the *Times*' senior-most political correspondent and a national treasure who'd been covering presidential campaigns since I was getting flicked in the head in the hallways of Hobby Middle School.

Carolyn is a master motivator. She knew all our pressure points and pressed them in exactly the right way to get us to perform at our highest levels even when we felt so out of fuel that we'd prefer to collapse on the field—as I did that day in Manchester. My pressure point was the fear that one of my more esteemed colleagues would overshadow me on the Hillary beat and prove The Guys right. I fell for this every time. Explain to Carolyn that I had nonrefundable theater tickets and needed to leave the office by 7:30 p.m.? "Well, I can see if Pat wants to write it." Start to spit out that Bobby planned a special anniversary dinner for us? "Okay, no problem, I mean, if you're busy, I can check with Michael." And every time, I replied just as I did by the indoor pool. "No, I'll do it."

"Fabulous! You're a star," Carolyn said, popping up from the lawn chair, its silver plastic strips sagging from where she sat.

HARRY REID ASKED if I was enjoying his state. "We've got no humidity here, so your hair won't frizz," he said. I realize this may seem like an odd thing for the senior-ranking Democratic senator to say, but my only thought was, how did he know? For months in Iowa and New Hampshire, my brittle split ends made my hair the consistency of a West Texas tumbleweed with the added static of my synthetic wool *Times* hat that I hardly ever took off except to sleep and shower. But thanks to a couple of days in the dry desert air of Las Vegas and the lemongrass citrus deep conditioner provided by the Wynn hotel and casino, my curls felt springy and light.

"As a matter of fact, Senator, my hair hasn't looked this good in months," I told him. "Maybe every Jewish girl should live in Nevada."

Reid was right. A day into my Nevada bureau chief assignment and I felt like the warden had urged me to orchestrate a prison break.

Three days earlier Bernie had trampled Hillary in New Hampshire. It was a clear, crushing, browbeating annihilation. Brooklyn knew Bernie would win by double digits, but they were thinking eleven or twelve points, not twenty-two fucking points.

Hillary thanked supporters at Southern New Hampshire University in Hooksett, in the same gym where eight years earlier she'd unexpectedly beaten Obama and declared she'd "found my own voice." I wrote that she left New Hampshire that night still searching for it. "Do we have any sense from her what she believes or wants her core message to be?" Joel Benenson had asked the team.

Before I could set up the *Times*' Nevada bureau (also known as room

1009 at the Wynn), I had to make a quick trip to Milwaukee for the next Democratic debate. I was in a Delta comfort seat when Brown Loafers came down the aisle and checked his boarding pass. He looked at me in the seat next to his, made one of his signature facial expressions—a feigned overbite, chin-forward grin—and exhaled as he sat, cursed by New Hampshire primary voters and the commercial airline Gods.

We made cordial small talk. I told him about the Iron Horse Hotel, my favorite boutique hotel in Milwaukee ("It's nicer than the Bowery and costs like a hundred and twenty dollars a night . . ."), and about the local coffee ("You have to try Colectivo . . ."). I said Hillary needed to win the primary so my colleague Mark Landler's upcoming book about her foreign policy would sell. "I've got a fair amount riding on this, too," Brown Loafers replied. Then he flipped through the *Times*. He pretended to read for the rest of the flight, thumbing well into the vitamin pages, what we call stories that get banished to the inside pages—B6, B12, and so on. I pretended to nap.

Because it would've been more awkward not to, we shared an Uber to the University of Wisconsin and as we rushed inside the auditorium with minutes to spare before the debate started, a protester shouted, "Hillary Clinton is a criminal!"

"AND YOU'RE AN ASSHOLE!" Brown Loafers yelled back.

The protester, sweeping his homemade sign down to his side, started to lunge toward us, yelling, "Hillary FOR PRISON!"

We sprinted up the stairs, past security, through steamy glass doors, locked to outsiders but open to us because we had Secret Service credentials. Through the glass, we heard yells of "Go FUUUUUUCK yourselves!" We both laughed—real laughter for once, not the scoffing, cruel kind that we'd directed at each other so many times. We shared something, an instant that I forgot about entirely until many months later when the assholes of Wisconsin delivered the election to Trump.

After the debate, Hillary added a speech in Harlem, a reset in the city that loved her the most. This gave me one night in New York, enough time to swap out my warmest winter clothes with some tank tops, sundresses, and a pair of sandals. Bobby and I tried to have breakfast at our favorite neighborhood diner, but I had to rush home to write a news story on Justice Antonin Scalia's death, leaving Bobby with a plate full of uneaten bacon and the bill. Later that day I took a one-way JetBlue flight to Las Vegas.

AS HILLARY MORPHED into a better version of herself in Clark County, so did I, letting go of all the crap that had built up in Iowa and New Hampshire. After so much time in states that felt alien, Las Vegas, especially away from the strip, felt like San Antonio. There were rows of almost identical stucco homes like the ones my parents downsized to a couple of years earlier, behind walled subdivisions with names like Alto Mesa and Hacienda Park. There were strip malls with tanning salons even though the temperatures in the winter rarely dipped below a sunny sixty-five degrees. The Mexican restaurants sold tacos *al pastor* and didn't charge for chips and salsa or try to top it all off with melted cheddar. And Spanish was everywhere, not like in Des Moines or Manchester where the language was treated like background noise from a kitchen.

In Vegas, Annie Karni and I took boxing lessons with Brady, a former Ultimate Fighter with a shaved head and a sinewy frame. Brady obviously hadn't met many journalists because he was shocked at our willingness to punch each other. I'd had so much pent-up rage and office drama playing out in my head that when Brady yelled "Left hook!" I'd thrown an uppercut that knocked his plastic name tag off.

That night Annie and I went to a Britney Spears concert. When a couple of male dancers in black leather first carried Britney onto the stage in an emerald-sequin-and-nude getup that hardly covered her gyrating ass, a man behind me said, "Good for Brits, she got her body back!" I wondered if I'd ever get my precampaign body back. A few songs in, while Brits flipped her blonde extensions as she lip-synched "Work Bitch," I took a swig of my seventeen-dollar margarita and realized that, for the first time in months, I wasn't thinking about Hillary Clinton.

By the time we got in a black car back to the Wynn, we were back to our usual stressing about newsroom dynamics and the campaign and our insecurities about what we and our colleagues would do with our lives when it all ended.

"What's going to happen to Maggie if Trump wins?" Annie asked.

The driver interrupted to say in a thick Bronx accent, "Yo, I don't know who this Maggie is, but what's going to happen to the country if Trump wins?"

Hours later I saw that Britney, all covered up in a tight white turtleneck and jeans, went to meet Hillary at her suite at Caesars, posting a photo, their heads of iconic bottle-blonde hair touching, on Instagram, and calling her "an inspiration and beautiful voice for women around the world!!!" When the campaign interpreted the post as an endorsement, Britney (or more likely, her legal guardians) deleted the #ImWithHer hashtag, causing a mini scandal and putting an inevitable Hillary taint on my memory of the Britney show.

AT THIS STAGE of the race during the '08 primary, I'd been the pool reporter assigned to "WJC duty." I'd never seen anything close to the tickled anarchy that ensued when Bill Clinton walked into a casino unannounced. *Three, two, one, chaos* as the cocktail waitresses nearly dropped their trays and the senior citizens who sat on padded swivel chairs shoving quarters into the slot machines broke away from the glowing red sevens and the tourists in town for a bachelor party seemed to yell, "Bro! It's Bill Clinton!" in unison.

"I'm from Arkansas, I met you in 1987 in Pine Bluff," a cook at Caesars Palace told the former president. "Yeah, Donna, I remember that. How's your momma doing?" I didn't see Clinton catch a passing glimpse at her name tag, but he must have.

We'd been in the basement of the MGM Grand with showgirls and kitchen workers lined up for their chance to shake his hand. Bill Clinton had turned and looked me in the eyes and said something that, eight years later, I kept going back to: "We had a little role reversal in New Hampshire. Hillary started running more of an insurgent campaign, an underdog campaign, and we like it that way. That's how it ought to be. We've been running as the underdog most of our lives."

Hillary arrived back in Nevada the underdog. A couple of months earlier Bernie had been a nonentity in the state, but now his ads blanketed the airwaves in English and Spanish. He had organizers from Elko to Carson City and polls showed he could edge out a victory. For the first time of the entire campaign, it dawned on the entire cocksure Clinton operation that she could actually blow this thing . . . again.

If there was any upside to the blowout in New Hampshire and the tightening polls in Nevada, it was that losing brought out Hillary's best self. I saw it over and over. Bill was right. She needed to run scared—as she had in 2008 after Iowa. It was the only way for her to tap into a deep reserve of retail-campaigning talent. Hillary couldn't truly pour herself into a room unless she felt as though it was all slipping away.

It was 12:35 a.m. local time on the Thursday before the Nevada caucuses when Hillary, in black slacks, glasses on, hair flattened from a day of travel, took the elevator downstairs at Caesars Palace. She headed toward the bustling heart of the hotel where mostly Latina housekeepers folded sheets and towels with tight corners and stacked them in bottom-heavy piles topped off with pillowcases and face towels.

"My goodness!" Hillary said, turning her head right and left to examine the scenes of curvy women in blue-and-black cotton uniforms loading their pillars of terry cloth and seven-hundred-thread-count cotton into wheeled plastic containers the color of an orange traffic cone. They lifted cupped hands to open mouths, blown away with the luminary in their midst. This wasn't Wayne Newton or the delicate acrobats in leotards and eastern European accents who wrapped their legs over their heads and contorted themselves into human bicycle wheels each night in the Cirque du Soleil, all of whom essentially belonged to the same casino ecosystem as the lower castes. This was Hillary Clinton and despite the drumbeat from Democrats that her campaign was joyless and a sentiment that even some supporters had summed up as "I'm With Her . . . I Guess," Hillary's impromptu visit—hair messy, makeup untouched up—moved these women, on hourly wages and tired feet, to tears.

"Ay, Dios mio!"

"Mira!"

"La Hillary!"

She reached both arms out in a wide embrace, one woman under each arm as if they were aunts lined up for a *Quinceañera* portrait around a fifteen-year-old in a tiara. They screeched and dug into deep pockets pulling out Android phones to snap photos. "I flew in from Chicago, so before I went to my room, I said, 'Well, who is still working?' The answer? A lot of people!" Hillary said as she made her way through the windowless room with its lemony, antiseptic scent of industrial-strength detergent. "Good to meet you. Thank you all. How's it going?"

She asked about the women's hours. The 5:00 p.m. to 1:00 a.m. shift, they replied, as they kept folding towels, afraid to get behind. Hillary asked what they did. "So, it's just towels and linens every day?" They chuckled politely at this question. Yes, towels and linens every day. For a second, I wondered if Hillary, who had lived her life tethered to the daily uncertainty of the inevitable next crisis, felt some longing for this tedious, predictable work. Towels and linens every day.

Hillary did this night after night before the Nevada caucuses. She met workers at the Paris Las Vegas Hotel and Casino ("How does it work?" she said to a woman who loaded dirtied sheets into an industrial washing machine.) and snuck up on a couple of tattooed cooks on break at Harrah's. At La Flor de Michoacan Ice Cream Shop in North Las Vegas, she eyed the Mexican flavors, tamarind and guava and chocolate. ("LIME!" Brown Loafers yelled when the Travelers asked what flavor ice cream she'd ordered. Hillary laughed.) And at a children's soccer game in Henderson, she pretended to be a goalie, shifting left and right on the synthetic grass and yelling, "Oh no, no, oh no!" as a runt of a six-year-old in a glowing yellow jersey kicked the ball through the nets. *Goal!*

She embraced all Nevada's eccentricities, including the endorsement of five hundred sex workers, mostly from Carson City brothels, who formed the "Hookers 4 Hillary" group. Even when there were only a couple of hours to go before the caucuses, Hillary tacked on a stop to Harrah's casino. "I need your help this morning, in the showroom at eleven a.m.," she told the Spanish-speaking hotel staff.

Hillary's style was so ripped from her husband's handbook, we might as well have been watching Bill Clinton work the crowd at the Bradley County Pink Tomato Festival in Arkansas. Despite what everyone (including Hillary) said about Hillary's political shortcomings, she dazzled at it . . . when she wanted to.

The weekend before the caucuses when a ten-year-old girl teared up and said she was scared that her parents would be deported, Hillary pulled her under an arm and held her close. "I'm going to do everything I can so you don't have to be scared and you don't have to worry about what happens to your mom, your dad, or someone else in your family," she said. "Let me do the worrying. I'll do all the worrying. Is that a deal?" The campaign turned the moment into a shaky-cam ad that ran throughout the primaries.

It was this scrappy, downtrodden Hillary that Democrats had in mind after she left the State Department, and every other capable Democrat including the sitting vice president had decided they didn't stand a chance against.

"She was a very good candidate in 2008 after she got knocked back," David Axelrod, who was no fan of the Clintons, had to concede. "Instead of a battleship, she became a speedboat, and she got down on the ground and really, I thought, really connected to the middle-class voters and people who were struggling. People who were struggling connected with her when she looked like she was struggling." But that Hillary had hardly shown herself on the campaign so far. That fall, the pollster Peter Hart concluded that Hillary appeared to be behind a "glass curtain." Many voters, he said, "feel they can see her and hear her, but they do not think they can relate to or touch her . . . In their words, she is distant and remote."

I asked Brooklyn why Hillary wasn't running like she did in 2008 when she was losing to Obama. Jen answered pointedly, "Because she's not losing."

29

"You Should Be So Pretty!"

LAS VEGAS, FEBRUARY 2016

Working on Pacific time meant New York only needed me at 5:00 a.m. PT when editors were heading into the office and usually wanted to know what the mood was on the ground.

"Are they nervous?"

"What are the Clinton people saying?"

"Do their internals show Bernie ahead?"

To which I replied, with varying degrees of politeness, "It's five a.m. here. Let me check on that when people wake up, and get right back to you," and went back to sleep until the sun came up over the Las Vegas Valley, turning my tenth-floor hotel room at the Wynn into an explosion of peachy nude tones. The tufted sofa and its ottoman, the floor-to-ceiling curtains that swung open at the flick of a bedside switch, the lampshades that glowed pink on top of marble bases, the circular table with the in-hotel magazines fanned out on top, the checkered carpet, and the leather office chair on its silver swivel base, all bathed in peach. This color scheme could only exist in a high-end hotel in a city where winter boots meant thigh-high Christian Louboutins. I squinted at the reflection that bounced off the gold edifice with the word TRUMP stamped in Chinese aluminum from the Trump hotel just outside my window. The *T* on the tower's white rim practically cast a Batman shadow on my California king bed. In the other direction stood the Treasure Island where I'd stayed with my Grandma Rose on a trip to Vegas for my twenty-first birthday.

Everything I saw in Vegas reminded me of my dad's mom, an artist and card shark who slept in silk negligees, traveled with a miniature wet bar, and had the misfortune of being born a world-weary woman in 1920 in a dry, heavily Baptist county in Waco, Texas. This affliction, of being trapped in an era before her time, gave Grandma an underdeveloped maternal instinct and almost nonexistent brain-to-mouth filter. When I was eleven, I propped my head up on bent elbows lying on the shag carpet in the living room of her pink ranch house. *Wheel of Fortune* was on, and I committed the unspeakable faux pas of saying I didn't think Vanna White was *that* pretty. Grandma nearly choked on her gin martini. "You should be so pretty!" she said from the sofa and then patted her lap as she yelled, "A Change of Heart!" completing the puzzle when the lineup of TV contestants were still mulling whether to buy another vowel.

Widowed twice, Grandma had, in her twilight years, a stash of untouched savings that provided her with a comfortable-middle-class life that included a Lexus and regular Southwest flights to Vegas. On my twenty-first birthday trip, I hardly saw her despite our sharing a room at the Treasure Island.

"You're asleep already?!" Grandma said when she turned on the lights in our room and rummaged around for her checkbook so she could get back down to the tables. On the last day of the trip, she woke me up early and offered to buy me a Bloody Mary at the tables. "Grandma, they're free," I said, pulling a pillow over my head.

Grandma Rose had been dead and buried for ten months when I felt her resurrected in Vegas. I couldn't walk by a blackjack table without thinking about Rose in the sequined boleros with matching earrings and handbags that she'd painted on with hot glue and glitter bought in bulk at the Hobby Lobby and that she wore daily in Waco but that really only made sense in Vegas.

I remembered the Treasure Island being swanky when we'd stayed there. But when Bobby, who came to visit me for a weekend during the pre–Nevada caucus stretch, and I went back to play the five-dollar blackjack tables, the lobby felt worn in, neglected. The casino reeked of cigarette smoke. The retail offerings included a Krispy Kreme, a Gilley's Trading Post with embellished Western-themed denim jackets, and an accessories store that sold rhinestone-studded purses and gold lamé pash-minas, all for under ten dollars. It was all so *Grandma*. We lost a hun-dred dollars in less than forty-five minutes and left cranky and swearing off the Treasure Island altogether.

By day three, I'd turned the word *Grandma* into an adjective I used so frequently it's shocking it didn't find its way into my *Times* copy. The Sinatra-themed Italian steakhouse Bobby and I ate at, with its red table-cloths and matching red-lacquered chairs and menu items like "Osso Buco My Way" wasn't bad, but it was definitely Grandma. On day nine, I came close to splurging on a pair of royal-blue satin flats with a mirrored, diamond-lined spinner affixed to the toe like a roulette wheel, displayed behind glass in an indoor walkway on the way to the Wynn's free parking garage. But were they too Grandma? I sent a picture to Michael Barbaro in the New York newsroom asking for his opinion. He wrote back imme-diately, "Come home. Right now. Your judgment has been compromised."

THERE IS A photo of Hillary, sandwiched between Grandma and me after a rally in Waco before the Texas primary in 2008. A week earlier a Dallas policeman, Victor Lozada-Tirado, a forty-nine-year-old father of four, was accompanying the Hillary motorcade when his motorcycle hit a curb and spun out of control. I was looking out the window when it happened. In an instant, the skid of a tire on the cement, his bike flipping behind him. The motorcade kept moving. By the time we got to the Dallas rally, we heard he'd been killed. Hillary said she was "greatly heartsick over the loss of life in the line of duty." The alt-right website Infowars added the officer to the list of "The Clinton Body-Count." A CNN reporter asked if Hillary would still be taking questions from the press. "Somebody just DIED!" Jamie, the '08 campaign press wrangler, yelled, the only time I'd ever heard her lose her temper.

Seeing someone die out the window of the bus doesn't go away. I was washing my hands in the bathroom and drying them with a soggy roll of stiff brown paper towels at the Waco event when I decided to do something for Grandma that I'd never done for anyone else in my family: I asked her if she wanted to meet Hillary after the rally. Grandma was reapplying her *Miami Vice*–red lipstick in the mirror when she replied, "And why would I want to do that?" I don't remember ever talking politics with Grandma. She wasn't one of my relatives who watched Fox News all day or had the car radio set to Rush Limbaugh. I don't remember her ever saying who she voted for or if she voted at all. I replied with the one reason I thought would sway Grandma: "Because she's the former first lady and is famous and everyone wants to meet her."

"Oh, all right," Grandma said, tossing her lipstick back into her handbag.

The second Hillary saw Grandma in her charcoal suit with black piping along the collar and an onyx pendant that matched her earrings, she said the only thing that could've turned Grandma into a lifelong Hillary devotee. "Oh, I love your suit! That's gorgeous," Hillary exclaimed before we crowded around her for a quick photo. Grandma melted. That photo booted pictures of all her other grandchildren and both of her deceased husbands off the TV mantel and gave me, the you-should-be-so-pretty granddaughter, the most prominent position in her house until she died. I have at least two chins in the picture and am stretching out a corduroy blazer to within an inch of its life. I'm standing to Hillary's left, lopsided, my shoulder weighted down by my suede leather satchel with my ThinkPad and my notebooks and my multiple water bottles. Grandma stands to the right, and in the middle is Hillary in a coral blouse and a matching coral choker (color coordination after Grandma Rose's heart). Hillary looks so young, a good twenty years younger than she did by the end of the 2016 campaign.

The day of Grandma's funeral, I was in the middle of closing a story and needed to explain why I'd be late with the fact-checking. Trying to buy time and a little sympathy, I told The Guys my Grandma died and sent the photo of the three of us in Waco in 2008.

A couple of weeks later in New York, I received a letter with a post-office box return address. I ripped it open, not thinking much of it.

Dear Amy:

I was so sorry to learn of the loss of your beloved grandmother, Rose; please accept my heartfelt condolences. At this difficult time, I hope you are comforted by the love she shared for you, the many lessons she imparted, and the memories you cherished together.

I am so grateful that I had the chance to meet your grandmother on the campaign trail in 2008; [The Guys] shared the lovely photo of us from that occasion. As you mourn her passing and celebrate her life, please know that my thoughts and prayers are with you, your family, and all of those whose lives she touched. Your grandmother was loved by many, and she will be dearly missed.

With deepest sympathy, I am.
Sincerely yours,
Hillary [signed the parallel scrawl of an ink-blue fountain pen]

30

Prince Harry

LAS VEGAS, FEBRUARY 20, 2016

Hillary needed to win Nevada—which by then I'd learned to always pronounce as "Nehv-ADD-uh," and never "Ne-VOD-a." Another loss would've crippled Brooklyn, already dumbfounded and dysfunctional after underestimating Bernie. ("He's not even a Democrat!" was a refrain from Clinton hands, almost as popular as, "He literally knows *nothing.*")

Robby's survival rested on Nevada, too. He'd been Hillary's state director there during the 2008 caucuses where he pulled out fifty-one percent of the vote establishing his prodigy status (and some rowdy nights and bottle service at a nightclub named the Foundation Room).

Back then, the Culinary Workers Union, the all-powerful labor union that represented sixty thousand casino and hotel workers, had endorsed Obama. Bill Clinton accused the union's leadership of pressuring its members ("the hardworking people who wash their sheets and cook their food") to caucus for Obama. "I haven't seen tactics like that in a decade," Bill told me at Caesars Palace. His proof? Chelsea. "My daughter is a scrupulously honest person," Clinton said, as if to acknowledge that he wasn't. "She heard them, she heard what they were saying." Chelsea stood at a safe distance shaking hands, seemingly oblivious to her father using her as a pawn in this latest political fight. A little later, when she noticed I was wearing a LasVegas.com lanyard, she introduced herself to me. "Hi, I'm Chelsea," she said, extending a hand. I introduced myself back, telling her I was a reporter with the *Wall Street Journal*. Awkward pause. "Oh, I thought you were with Vegas.com," she said, and abruptly turned her back. That was the first time I met Chelsea Clinton.

Eight years later, Hillary had the Democratic Establishment on her side.

For months, Robby assured supporters that Nevada would be a "Western firewall." The diverse population reflected the country's changing demographics (i.e., young, Latino, Asian American) and would install Hillary in her rightful place as nominee in waiting.

But after New Hampshire, panic set in. Bernie, now flush with cash from online donations from nineteen-year-old Bros in boxer shorts, had gained ground in Nevada. Suddenly, the state wasn't diverse. On conference calls, Robby warned donors that Hillary could lose. Hired Gun Guy explained that "about eighty percent of caucus voters in that state are white," giving Bernie the same advantage he had in Iowa and New Hampshire. In addition to this being a significant bending of reality (Iowa and New Hampshire are 86.2 percent and 90.8 percent white, respectively, and Nevada is 49.9 percent white) the campaign's newfound portrayal of Nevada enraged Harry Reid, the most powerful Democrat in the Senate and arguably the most powerful man in Nehv-ADD-uh.

A decade earlier, Reid had lobbied Congress to give the state an early caucus precisely because of its diverse and booming Latino and Asian populations. Reid hadn't endorsed either candidate in the 2016 primary, and he was exasperated with the Clintons, but he also wanted his beloved caucuses, still treated by the political elites as the ugly stepchild of the early-voting process (especially that year when the GOP primary in South Carolina attracted the media's nearly undivided attention), to play a major part in choosing the nominee. And Reid, like most of the Democratic Establishment, worried that if Bernie continued to gain momentum, Hillary, who was their only shot at defeating the Republicans, would emerge from the primaries as damaged goods.

So on the Thursday morning before the caucuses, Harry Reid, who had publicly remained neutral, made a phone call that tipped the Democratic primary in Hillary's favor.

After some small talk about my hair, Reid told me he'd called D. Taylor, the president of the parent union of the Culinary Workers. The union hadn't endorsed either candidate and hadn't signaled whether it would push to get its members—housekeepers and cocktail waitresses and short-order cooks who heavily favored Hillary—paid time off on a busy Saturday afternoon to participate in the time-consuming caucuses. "He's been extremely cooperative," Reid told me of his talk with D. Taylor. "Probably a hundred organizers will be at the caucus sites and in hotels to make sure people know what they're doing."

And they did.

Dozens of women in chocolate-brown uniforms took a break from their jobs as casino porters at the Bellagio to line up outside the Milano I ballroom on the promenade level of Caesars Palace. They told me they were from Honduras and Nicaragua and Guatemala and Mexico. They carried boxed lunches in one hand and American flags in the other. All of them said they were there to caucus "*por La Hillary.*"

"She'll change immigration. She'll change the economy. She'll change *todo*!" said Dora Gonzalez, a fifty-four-year-old Honduran with bright-red lips and black waves of hair that fell to her mid-chest, held back with two silver barrettes. "And she's a *mujer*!" added her friend Elba Piñera, who wore gold hoop earrings and had reading glasses perched in her shirt collar.

The Milano I ballroom was coated in gold paint and gold Grecian-themed carpet and coral-colored walls. Elaborate three-pronged sconces filled the windowless room with amber light. The casino workers rushed to claim scuffed conference-room seats, set up in tight rows across the two-thousand-square-foot ballroom. Even some of the Latino men, cooks and security guards and busboys, waved signs that said WOMEN FOR HILLARY.

When a Spanish-speaking precinct captain instructed the crowd to sit on "*la izquierda*" if they were for Bernie, and "*la derecha*" if they were for Hillary, the Hillary side of the room erupted, thrusting their arms in the air, chanting "Hill-AH-ree! Hill-AH-ree!" This relegated to a sad little corner of the left side of the Milano I ballroom a handful of white men, some of whom appeared to be pit bosses, the buttons on their black vests temporarily undone, and a dozen or so younger Latino men, who held Bernie signs with the silhouette of his white hair and the words *Educacion Gratis!*

The caucus had the feel of a frenzied, romantic, proletariat coup unfolding in a midsize casino ballroom instead of the Plaza de Armas of some Latin American capital, and lasted no more than five minutes. Similar pro-Hillary caucuses played out at casinos across the Strip. But that wasn't the whole story.

The Culinary Workers Union's rank and file who lined up at the Wynn and Harrah's and the Rio and New York, New York and handed the caucuses to Hillary had only been there at all because their bosses had given them their blessings and boxed lunches. And that had only happened because the Democratic Establishment made sure that it would. The man known as Prince Harry picked up the phone and called this city's ruling class. He told the union bosses and hotel owners in so many words, *Nice casino you've got there. It'd be a shame if anything happened to it.*

When it was all over, Clark County, which includes Las Vegas and 72 percent of the state's population, made Hillary the winner by more than five points. She'd regained the momentum and headed into South Carolina on an entirely different trajectory than just hours earlier.

We started to write Bernie's obituary. Hillary devoted her win to "hotel and casino workers who never wavered" and "tens of thousands of men and women with kids to raise, bills to pay, and dreams that won't die." She used plural pronouns like *we* and *us* more than usual, avoiding *I*. "I'm on my way to Texas. Bill is on his way to Colorado. The fight goes on! The future is within our grasp!"

On deadline and with no office center in sight, I took the escalators two flights down to the casino where I spotted an empty electrical outlet next to a *Sex and the City*–themed slot machine and sat down to charge my laptop. A security guard came by and told me I couldn't sit there unless I was gambling so I loaded a few quarters and dollar bills into the slot machine as I typed, wondering if I could submit the expenses to our draconian newsroom administrator. Were they "Entertainment" or "Water, Snacks and Incidentals"? I walked away with forty-five dollars when Mr. Big aligned with two pairs of Carrie's Manolo Blahniks. A waitress dropped off a bottle of Caesars-branded water and I broke open the plastic cap and took a long sip. I looked at my phone and saw the news that Jeb! had dropped out.

Jen said it was the campaign's best day yet. Backstage, Hillary cried. On the way to the Elite Travel private terminal of McCarran International Airport, the Travelers and staff picked up six packs of Peroni, and everyone landed in Houston for a late-night rally feeling happy drunk.

31

The Plane Situation

MARCH 2016

The Friday morning before the South Carolina primary, Hillary made an unannounced stop at the Saffron Cafe & Bakery in Charleston. Still high from Vegas, she floated over to a table full of Southern frat boys, a groom and his ten groomsmen, each more clean-cut than the next. "That looks like a really good Bloody Mary!" Hillary said, eyeing their cocktails at approximately 10:44 a.m.

Barb Kinney, the campaign photographer, instructed the dudes to kneel down on one knee as if they were proposing and positioned Hillary at the center. "I loooooooove having men at my feet," Hillary said. The next day, she beat Bernie by forty-seven points in South Carolina.

Saturday, 2/27
　　Charleston, SC en route Birmingham, AL
　　Birmingham, AL en route Columbia, SC
　　RON [remaining overnight]–Columbia, SC

Sunday, 2/28
　　Columbia, SC en route Memphis, TN
　　Memphis, TN en route Nashville, TN
　　Nashville, TN en route Pine Bluff, AR
　　RON–Pine Bluff, AR

Monday, 2/29
Pine Bluff, AR en route Springfield, MA
Springfield, MA en route Boston
Boston en route Norfolk, VA
RON–Norfolk, VA

Tuesday, 3/01
Norfolk, VA en route IAD
IAD en route Miami, FL
RON–Miami, FL or commercial to YOUR final destination

With Trump surging, Republicans delivering bleary-eyed concession speeches almost daily, and Bernie appearing to be the walking dead, my editors had virtually forgotten I (or Hillary) existed as Super Tuesday approached. My colleagues on the investigative team had a gripping deep-dive into Hillary's legacy in Libya. The *Times*, afraid the story could affect the race, decided not to publish it until after the polls closed in South Carolina. I rolled my eyes when I heard this. I'd spent months talking to voters in black churches and cafés and union halls, and the last thing on their minds was Hillary's use of smart power in the ouster of Qaddafi. I received a late-night email from a Very Senior Editor who'd never commented on my coverage before. He asked me what the campaign's reaction to the Libya story was. "I haven't heard anything," I replied. "They're all pretty elated after last night's win."

I didn't know why the email upset me so much. Yes, I was hurt. I was in the midst of covering an evening rally in Pine Bluff, Arkansas, a time-worn, predominantly black town on the Delta Lowlands with a poverty rate more than twice the national average. I'd been on the road for weeks, had just logged another twenty-hour day, and wanted Very Senior Editor to acknowledge my work. The Oscars were happening during the rally, and we snuck peeks at Twitter in between transcribing Hillary's comments about Empowerment Zones and the New Markets Tax Credit and statistics from the Clinton years ("During Bill's terms, the median African

American family income went up thirty-three percent . . ."). New York and the *Times* newsroom existed on a planet as foreign to Pine Bluff as the red carpet at the Dolby Theatre. One where people had an entirely different perspective on the election than in the places I visited. They had different concerns from the people Hillary spent her days talking to, different expectations from the families who stuck TRUMP yard signs into their freshly mowed grass.

I couldn't answer Very Senior Editor's question even if I'd wanted to. Traveling *with* the campaign meant I knew far less about what was happening *inside* the campaign than if I'd been back in New York working the phones or meeting sources in Brooklyn.

It had been eighty-four days since Hillary held a press conference. We all bitched incessantly that we needed more access, but Hillary didn't see any reason to talk to us. With everything going her way post-Vegas, her media strategy could continue to consist of a couple of sit-downs with the celebrity talk-show host Amanda de Cadenet and GloZell Green, the YouTube comedian with the silvery emerald lips. Compared to the Trump press corps, whom he publicly humiliated and who had to dodge loogies and abuse from enraged Everydays (at least one of whom wore a T-shirt that said ROPE. TREE. JOURNALIST. SOME ASSEMBLY REQUIRED), the Hillary press corps was coddled. We didn't need bodyguards to escort us into her rallies, and while being ignored was unpleasant, it was nothing compared to a candidate who threatened to blow up the First Amendment.

Still, I would've endured being called third rate at a rally in Youngstown if it meant Hillary gave us as many interviews as Trump gave reporters. He put on a big show of hating us, but he'd get on the phone, no problem.

In a way, Trump's mistreatment of the media had done Hillary a favor by freeing her of the decorum of a traditional campaign. But it also meant the reporters who spent their days trying to cover and explain Hillary to the American public never got to bridge, as one reporter who traveled with the first lady in the 1990s put it, the "disconnect between the kind of person you could convey or are in private and amongst us on these trips, so much sense of humor, very warm and engaging in what we see on television or in the news." How could we communicate Hillary's "funny, wicked, and wacky" side to voters if we never saw it for ourselves?

Had this been any other election year, the candidate and top aides would be flying on the same plane as the traveling press as had been the practice in every modern presidential campaign since the dawn of mass air travel. Instead Brooklyn organized a separate press charter for the Travelers and billed our news outlets, while Hillary and her staff flew on their own private plane.

IN 2008, THE close quarters of Hill Force One and, later, the Change We Can Believe In Express led to the only real interactions I had with Hillary and Obama. I remember Hillary taking over the intercom to welcome us on the rickety 737. "Welcome aboard the maiden flight of Hill Force One. My name is Hillary, and I am so pleased to have most of you on board," she said. "FAA regulations prohibit the use of any cell phones, BlackBerries, or wireless devices that may be used to transmit a negative story about me."

We developed such solidarity on the shared plane that the press decorated our quarters with a Hillary Pantsuit Schedule, an oversize wall calendar with a rotating cast of Hillary in forsythia yellow (Monday), turquoise blue (Tuesday), fiesta pink (Wednesday), sunset orange (Thursday), show-stopper red (Friday).

The night of the Ohio and Texas primaries, the campaign plane hit gale-force winds, and I saw lightning flash against the wing tips. Tina Brown looked as if she was about to puke. I grasped hands with one of the New York tabloid reporters and, thinking we'd die, actually said, "Now, we'll never know if she won Ohio."

Peter Nicholas, who was with the *Los Angeles Times* then, loves to tell the story of the time Hillary walked in on me taking a piss. "She opens the door and there's Amy with her pants around her ankles . . ." he said to a small crowd of senior campaign aides and TV correspondents during the spaghetti dinner at Podesta's house. Peter always told the story with his usual theatrical peculiarity and amplified fidgets and grins, glances at me, at the carpet, me, the carpet.

I remember the moment exactly, when the accordion door with the broken lock began to split in its middle and fold into the campaign plane's back lavatory, the smell of the blue ammonium swirling in the bowl, a sunset-orange pant leg forcing its way in. I knew that hue didn't belong to the other reporters on the plane. An instant later it hit me: That was Thursday's pantsuit. *Hillary Clinton needs to pee.* I'd already pulled my jeans up to the fleshy part of my sit bones when we made eye contact.

"Uh, um, Senator, I'm so sorry, I guess the lock isn't working, and . . ." I started. Her facial features widened, and before she could say anything, a Secret Service agent inserted his tree trunk of a torso between us and Hillary backed up like Ginger Rogers sashaying into a rent-a-jet's service area. I rushed back to my seat with my fly still down.

Peter, who sat in the same row as me in the designated Major Dailies seats of the plane, noticed the commotion and asked in his loudest outdoor voice, "Um, did Hillary just walk in on you urinating?"

In April, when it was obvious Hillary wouldn't be the nominee, I switched over to covering Obama and assumed a seat on the Change We Can Believe In Express.

Two months before the '08 general election, Bobby surprised me and proposed. He had a sunburned nose and hair so dark it blended in with the night sky. He got down on one knee at the front of the bow on a sailboat in the Sea of Cortez and pulled the navy-blue velvet box out of the pocket of his cargo shorts, where he'd hidden the ring when we passed through immigration in Mexico a couple of days earlier. "You know the way I've always liked you . . ." Bobby said, over the yelps of sea lions and the waves brushing against the starboard. The air smelled of salt water and the green enchiladas cooking in the kitchen below.

I'd never seen myself as somebody's wife, but Bobby understood me better than anyone. He'd put up with long distance in Japan. When I left him again to meet up with Hillary in '07, he'd rented a car to meet me in New Hampshire. He'd built us bookshelves with the crafty optimism that we'd one day settle down. Or not. Either way, he knew what he was getting, and he'd always liked me anyway. And I'd come to think of him as my home, a safe, warm space in the form of a hunky Irishman whose very being told me everything would be okay and without whom I'd cease to exist. I said yes. Thirty-six hours later, I was back on the road in Obama's traveling press in Harrisburg, Pennsylvania.

It didn't take long before my trail friends gushed over my new radiant-cut diamond that still felt odd on my finger. Obama strolled down the aisle to the back of the plane where we were sitting. We rushed to ready our voice recorders and cameras assuming he'd do an impromptu press avail. Instead, Obama walked right up to me and said, "Okay, lemme see the rock." I stretched my hand out, and he entered dad mode. "That's great, so when are you quittin'?" He asked how long we'd been dating. ("Okay, three years is a decent stint.") He said he wanted to make sure I had plenty of time to plan the wedding ("After the campaign? Okay, well you know these things take time. You've gotta pick your dress . . ."). He offered to have the Secret Service and FBI do a "full background check" on Bobby. "Happy to. The FBI can get involved."

Bobby and I got married a year later in San Miguel de Allende, a colonial town in central Mexico. We had mariachis in white *charro* suits, a taco truck, and a donkey named Domingo that served tequila on his back. Our wedding programs included the Yeats quote, "Being Irish, he had an abiding sense of tragedy, which sustained him through temporary periods of joy." At the rehearsal dinner, my parents played the video of Obama congratulating me.

Eight years later, those kinds of intimacies were unimaginable.

EVERYONE BLAMED ROBBY for what became known as the Plane Situation. It wasn't true. Robby was a miser, but flying Hillary around on her own plane didn't save money. The campaign spent $2 million on private planes between June and January when the election still hinged on only a handful of states. It cost a fortune to maintain multiple private planes for "all three Clintons," not to mention the logistics involved. A refurbished Boeing 737 would've been more economical. The truth was Hillary didn't want to fly on the same plane as us. And because she was running against Trump, she didn't have to. He'd tossed all traditional decorum out the window of his twenty-five-year-old gold-plated 757—which, in classic smoke-and-mirrors fashion, he'd acquired from a now-defunct low-cost Mexican airline. Why couldn't Hillary have her own plane, too?

What Hillary didn't understand—or didn't want to accept—was that Trump could lick his fingers after eating a bucket of greasy KFC on board his 757 and maintain the aura of the workingman—even if he did spend most of his business career screwing over said working-man. Meanwhile, Hillary, who regardless of what you thought of her personally had detailed policy plans and a real determination to "lift middle-class wages" and "put Americans back to work," couldn't shake her image as a rich lady from Westchester. She flew on private planes and wore tunics that Everydays couldn't buy off the rack at a Macy's. Hillary's girlfriends finally forced her to stop saying, "You shouldn't have to be the granddaughter of a president or a secretary of state to receive . . . all the support and advantages that will one day lead to a good job and a successful life." Hillary loved the line, but it only reminded voters that she was one-half of a political dynasty. At one event, when Hillary did the granddaughter line ("Eighty percent of a child's brain is formed before the age of three . . .") a longtime aide leaned over and whispered to me, "I guess I was brain-dead by age four."

But nothing made Hillary seem like more of a monarch than her insistence that she be cocooned in the clouds at thirty-six thousand feet a safe distance from the press and the Everydays, surrounded only by her royal court, security, and a spread of crudités. And as the campaign's true power center—meaning Huma, Brown Loafers, Jen Palmieri, Policy Guy and Dan Schwerin (who both traveled on occasion), and assorted friends from the White House years—flew around in a pimped-out fuselage, it became easier for Hillary to ice out Brooklyn. Several senior aides thought she should share a plane with the press, the way she had in 2008 and at the State Department, arguing that the close quarters always led to warmer coverage. "She doesn't want to do it," Huma would say. Oh, and sorry, but there's no room on the plane for you. Betty Currie, Bill Clinton's secretary in the White House, once called Hillaryland "a little island unto themselves." In 2016, the little island had become a Falcon 900B heavy jet.

BEING ENSCONCED IN our own hulk of cherrywood and leather seats and cream-colored carpet in the sky meant the only regular contact the Travelers had with anyone outside our own ranks was the young press advance staffers tasked with keeping us on schedule. This was an almost impossible feat. We always stayed behind to pee or file our story or both. We made the bus wait if we forgot our iPhones in a bathroom stall (guilty) or got caught up in conversations with the Everydays. Now that we flew to multiple states a day, falling behind schedule meant the entire press corps could spend thousands of dollars to fly to the next city or state only to miss Hillary's speech. This may come as a shock, but neither she nor Trump waited for the traveling press to start.

These twenty-something press aides reported to Brown Loafers. I'd seen him as OG's marionette for months, but these aspiring press secretaries looked up to him as if he were the CEO of publicly traded spin. They took their main task—taking our meal orders—so seriously that I can only hope they'll one day be running the country. (Obama's press luggage handler in 2008, Eric Lesser, kept spreadsheets of our bags and once chased a red negligee rumored, incorrectly, to belong to Maureen Dowd down the tarmac. He is now a state senator in Massachusetts.)

"Below please find today's meal options for both lunch and dinner. Please send me your **lunch orders by 11:30 a.m.** and your **dinner orders by 4:00 p.m.**," they'd email, usually attaching a Panera Bread menu that I'd memorized a couple of states back. We'd often mull the menu and respond with two words: Hard Choices. When one staffer sent us a stern reminder the night before "Please reply to me with your order by **7:00 p.m. (EST) today**" for the following day's lunch order, we wrote back, "This is getting ridiculous," and "TOO SOON." They'd send around alphabetized Excel spreadsheets . . .

Alba, Carmen	Curried Cauliflower and Kale Salad
Becker, Amanda	Curried Cauliflower and Kale Salad
Chozick, Amy	Quinoa Salad with Chicken (or Turkey)
Earle, Geoffrey	Pistachio Chicken Salad Sandwich on Multigrain
Epstein, Jennifer	Curried Cauliflower and Kale Salad with Chicken
Fraser-Chanpong, Hannah	Chopped Salad with Chicken

. . . so that by mid-March we all knew each other's dietary preferences. NPR's Tamara Keith doesn't eat cheese. Reuters's Amanda Becker is a vegetarian. Annie Karni once threw off the entire spreadsheet system by sending in a late change to her lunch order (from salmon to the barbecue chicken plate). "Annie, please update me on the snacks you plan to have on the plane," Dan Merica emailed the group. Ruby Cramer wrote, "Really hope the BBQ was the right choice after all."

After days of nonstop bitching, The Guys decided to throw us Tony Goldwyn. I would've preferred if Goldwyn remained the villain from *Ghost* who came into my life in 1990 at the same time as the Righteous Brothers and Patrick Swayze's abs. Instead, the Hollywood actor became the embodiment of Hillary's neutered (or should I say spayed?) traveling press.

We were in Nashville when the actor sat with his legs spread wide across one of the leather seats of our luxury bus rental and gushed about how much he *loved* campaigning for Hillary. He may or may not have known The Guys were using him as a pawn. I stood up in my seat, my back pressed against the window to get a profile shot of Tony, his salt-and-pepper waves, the first few buttons of his white dress shirt left undone under a navy-blue blazer. By now everyone knew Tony as the philandering forty-fourth president who governed, for the most part, on his ability to fill out a tailored suit and squeeze in a West Wing quickie. His role on the soapy network drama *Scandal* gave his ubiquitous presence on the campaign a certain meta quality that Hillary ate up for the same reasons she loved to drop that she binge-watched *The Good Wife* and *Madam Secretary* and *House of Cards*—all TV series that ranged from exhibiting casual Clinton undertones to small-screen propaganda.

But mostly, Hillary loved Tony because he was hot and devoted to her. His jaw was shaped like a refrigerator, and he had feral gray eyes, and beneath all of that was an easiness and intellect that caused the women of Hillary's press corps to abandon whatever story we were working on to flip our hair and ask useless questions like "What did you think of Iowa?" and "How does Hillary seem today?" The Guys knew us so well.

Later that afternoon, I lost any shred of self-respect I had left when Tony broke the only news we'd gotten out of Hillary in weeks. She was shaking hands with college students at the Fido coffee shop in Nashville, a converted pet shop with exposed brick walls and track lighting, when Goldwyn asked what she thought of Trump not denouncing David Duke's support (a question she'd ignored when the Travelers asked).

"Oh, that's pathetic," Hillary said in Tony's direction, a response we overheard and all used in our stories. The CNN chyron flashed HILLARY CALLS TRUMP ON KKK LEADER "PATHETIC."

That's when Huma handed Hillary a latte in a to-go cup, made with locally grown Tennessee honey, cinnamon, milk, and espresso, and said to her boss, "They're amazing. They're making a ton of them for us to take." The Travelers watched this, our mouths watering, as Hillary and Huma headed out a back door.

32

The Gaffe Tour

MARCH 2016

The less I interacted with Hillary, the greater her imperial hold on my brain became. I used to get my hair blown out straight so I could go a week (nine days if I didn't sweat) on the road without washing it. But in early March I quit doing this, afraid Hillary wouldn't recognize me if she looked out at the press scrum, decided to acknowledge us, and didn't see my usual mane of brown curls. I started to have nightly Hillary dreams. We usually gossiped like old girlfriends. We were trying on clothes in adjacent dressing rooms at a Zara or maybe it was a Mango, but it was definitely in Barcelona. I couldn't button a size six pair of pants, and Hillary told me I must be pregnant. I said it was campaign weight, and we argued like that from underneath the dressing room stall for what felt like hours. I'd wake up and relay these dreams to Bobby, until even he started to question my sanity. "Fucking hell, with the Hillary dreams," he said, rolling onto his side away from me one Sunday morning.

Hillary won eight of the thirteen contests on Super Tuesday. We'd entered the "Bernie, Who?" phase of the primary when she'd refer only vaguely to "my esteemed opponent." Hillary had been so pumped after South Carolina that she'd even showed flashes of Saint Hillary, preaching about the country's need for more "love and kindness." At a rally in Nashville, when she heard a chant of "We love you Hillary!" she stopped and said, "You know, I'm all about love and kindness, so I sure appreciate that." Her reworked stump speech focused almost entirely on denouncing Trump. The most controversial woman in American politics for the past couple decades had finally found her raison d'être—the Great Unifier.

"What a Super Tuesday!" Hillary said at her victory rally at the Ice Palace Film Studios in Miami. "We know we've got work to do. But that work, that work is not to make America great again. America has never stopped being great. We have to make America whole—we have to fill in what's been hollowed out."

Trump was no longer the Pied Piper. He was the unacceptable alternative who would motivate Democrats and drive independents, moderate Republicans, the half of the country who disliked and distrusted Hillary, to Her side. I have to admit, I liked the retooled message. Hillary still had no clear rationale for why she wanted to be president. But reminding voters that she wasn't Trump seemed like a better reason than building ladders of opportunity.

But even as Hillary declared in Miami that "trying to divide America between us and them is wrong, and we're not going to let it work," back in Brooklyn, Robby et al. were feverishly slicing and dicing the nation's demographics right down to zip codes and church attendance and TV-viewing habits. There were churchgoing black women who loved Hillary ("She was loyal to her husband and she'll be loyal to us," one woman at the Mt. Zion Fellowship church in Cleveland told me).

The so-called Southern firewall was working. Hillary ran up the score with black voters in Alabama, Georgia, Tennessee, and Virginia and Latinos in Texas. Bernie won whites in Oklahoma, Vermont, Colorado, and Minnesota. Brooklyn started to worry black voters *might* feel exploited. After Super Tuesday, Brooklyn raced to install "a black campaign vice chair or Sr advisor" to "send the message that, Hillary puts her actions where her mouth is, and actually does appreciate the black vote." Enter Minyon Moore, a powerhouse and Hillaryland alum. Like most of Hillary's female friends (many of whom were women of color) Minyon was equal parts terrifying and a lot of fun. I'd heard she defended me early on after The Guys portrayed me as a Lucifer in Lululemon. She said to Hillary, "Are you kidding me? She's just a young woman trying to do her job." Minyon set up her office on the tenth floor of the Brooklyn HQ near Podesta, Mook, Jen, and Policy Guy. But for many young black voters, the stench of exploitation remained, even months later when Obama said he'd consider it a "personal insult" if black voters didn't support Hillary. "Absolutely hate this framing of the necessity of voting (for Democrats) that places all the blame on marginalized and young folks," Vann R. Newkirk II, a writer for the *Atlantic*, tweeted.

Hillary was baffled when everyone said she'd been pandering to black voters when she told Power 105.1's *The Breakfast Club* that she carries hot sauce in her bag. She's always traveled with hot peppers and hot sauce. And when she heard about the lyrics to Beyoncé's "Formation" off of *Lemonade*, Hillary initially thought the singer put hot sauce in her lemonade. "Who was I pandering to? The hot sauce lobby?" she'd ask people.

HILLARY HAD HARDLY left the stage at the Miami Ice Palace when Robby sent out a memo essentially predicting her defeat in Michigan. He noted that Bernie had spent $3 million on TV ads in the state and is "competing very aggressively in Michigan." Robby and his delegate calculator didn't see the point of investing much there—the math didn't make sense. "Even if Sen. Sanders were able to eke out a victory there, we would still net more delegates in Mississippi, which holds its election on the same night."

Who needs Michigan when you've got Mississippi?

I figured Robby was downplaying expectations. The campaign's internal polling had Hillary at least five points ahead in Michigan. I was so certain she'd win that the night of the primary I decided to skip her rally in Cleveland. I wasn't writing the main news story and getting there would've involved a four-hour drive from Detroit and an early-morning flight to Miami. But those were excuses. The truth was I wanted to see Bobby and sleep in my own bed. I could live-stream the speech.

But as results trickled in, my editors panicked. "You get any sense of the mood over there?" Carolyn emailed. I didn't lie. But I also didn't volunteer that I was on the Lower East Side. Mamma wouldn't be happy if she knew I was at home watching Wolf Blitzer while slurping down tiny buns in my pajamas. "It's not good," I emailed. "She has got to kill this narrative that she is a 'regional' candidate who only wins in the South."

Both of these statements were technically true: The mood in our living room that night was definitely not good, and Hillary had to prove she could win whites in the Midwest. Aside from worrying that I'd let Carolyn down, my main thought when CNN called the contest for her Esteemed Opponent was *Fuuuuuuuuuuuuck, this motherfucking primary is never going to fucking end.* I stomped around the apartment making calls to sources about why she lost and what it means. I booked my flight to Miami in the morning and threw my roller bag open on the bed and stuffed it with layers to get me through several climates, while reciting something like "Fucking, motherfucking Michigan. Fuck. Fuck." By midnight, Bobby wished he were watching CNN's magic map all by himself.

In her Cleveland speech, Hillary didn't mention losing Michigan or winning Mississippi, but for the first time she showed the subtlest of public signs of what she'd been saying privately for months—her lifelong belief that we're all in this together just didn't fit with the off-with-their-heads election year. For eleven months, I'd heard Hillary say, "I want to be the president for the struggling, the striving, and the successful." But that night she said, "I don't want to be the president for those who are already successful—they don't need me. I want to be the president for the struggling and the striving."

Hillary was on a rampage. Several Brooklyn sources told me she'd been as exasperated by my Michigan postmortem and the *Times'* coverage as she was by Bernie Sanders, and that was saying something. Hillary may have even preferred stories about her emails or her psychological state to stories about her evolving message (See "Spontaneity Is Embargoed Until 4:00 p.m."). She thought I'd blown up a slight variation on her standard line into the front-page, above-the-fold story that said she'd been "stung by the bad showing" and "was already recalibrating her message, even altering her standard line before the Michigan race had been called." I wrote that the state's white voters had been "scarred by the free trade deals" like NAFTA that Bill Clinton signed into law.

I know this irked her, but honestly, I couldn't believe Hillary still hadn't figured out what to say about NAFTA. During the 2008 primary, she bashed her husband's signature trade deal in Ohio. Then we'd fly to south Texas—where the economy was going gangbusters thanks partly to NAFTA—and she'd praise its impact. Obama summed it up in 2008: "The fact is, she was saying great things about NAFTA until she was running for president." Now Bernie talked nonstop about her paid speeches to Wall Street ("I kind of think if you're going to be paid $225,000 for a speech, it must be a fantastic speech, a brilliant speech which you would want to share with the American people") and her support for "the disastrous trade agreements written by corporate America."

THE NIGHT AFTER Michigan, at the next Democratic debate in Miami, Joel Benenson insisted Rust Belt voters didn't give a shit about NAFTA. They'd crunched the data and said trade didn't motivate the suburban women and minorities of the vaunted Hillary Coalition. "It's just not an issue that we're seeing," he said. Jen Palmieri shook her head at me outside in the swampy south Florida air. "NAFTA, Amy, really?" The next morning, after the Travelers flew from Miami to Tampa only to miss most of Hillary's speech, I complained via email, "We invest a lot of time and resources into traveling and expect, at the very, very least, to be at the events."

Brown Loafers replied, "Yes, I wouldn't want her message being mischaracterized." And, "I'm sorry you guys are delayed. But I have sympathy for just about everyone on the road but you today."

Thursday, 3/10
 Miami, FL en route Tampa, FL
 Tampa, FL en route Durham, NC
 Durham, NC en route Vernon Hills, IL
 RON–Chicago, IL

Friday, 3/11
 Press should make their own way to St. Louis, MO to resume traveling with us the morning of Saturday (3/12).

Saturday, 3/12

Press will meet at first location in St. Louis, MO. *Exact timing and location TBD* but will be sent to you as soon as possible.

St. Louis, MO en route Cleveland, OH

RON–Cleveland, OH

To make matters worse, I couldn't stop itching. I was convinced I'd picked up scabies at a La Quinta Inn at the Miami airport where a working girl in a tiny miniskirt had paid cash for her room before I approached the front desk to inquire about our corporate AmEx rate. I couldn't sleep, certain there was something alive and burrowing under my skin. I couldn't get out of the bubble to see a doctor until a couple of days later when the campaign got to Chicago and Hillary took a day off to attend Nancy Reagan's funeral in Simi Valley, California. For two days, I tried to type with one hand so I could scratch with the other. My arms and legs were dry and dotted with puffy red marks.

"Not scabies," the doctor said, as he scribbled on a prescription pad. He said the agonizing itch was an allergic reaction to cheap hotel detergent and gave me a prescription for an antihistamine and an ointment that made my suitcase smell like sulfur.

I wasn't convinced. "Are you sure? I texted my friend Barry who got scabies from the towels at our boot camp and he said it looks like scabies."

He was losing his patience. "Mites are parasites. They would've nested and bred between your fingers. There's nothing there."

"Oh, thank God, thank God, thank God. That seemed disgusting and awful, and I definitely wasn't gonna have time to fumigate my clothes before Tuesday."

I'd hardly jumped off the exam table when I looked at Twitter and saw Hillary had a momentary brain fart. In an interview with Andrea Mitchell she praised Nancy Reagan's "low-key advocacy" on HIV/AIDS. She said the first lady (who notoriously ignored HIV/AIDS even as thousands of gay men died) had "started a national conversation, when before nobody would talk about it." I'm away from my laptop for ten minutes, and she had to piss off the gays.

Hillary so rarely misspoke that every gaffe, if left hanging there, could stretch out for months, years, hell, her whole career (see Cookies, Teas). But the pressure to be perfect started to take on a whole new meaning when it looked like she'd be running against Trump.

"He can say whatever the fuck he wants, but she's exhausted and tried to think of something nice to say about a dead lady?" Jen asked me that night. Fair question. I didn't have an answer. I still don't.

(That night when Trump canceled his Chicago rally amid violent protests, I heard from Trump sources who said Brooklyn had staged the unrest to draw attention away from the Nancy Reagan AIDS gaffe. Again, giving the Clinton campaign way too much credit.)

The Gaffe Tour continued. Two days later, Hillary said during a CNN town hall in Columbus that her clean-energy plan would "put a lot of coal miners and coal companies out of business." She'd already hemorrhaged white working-class voters to the point that no one remembered that those had been her people in 2008. And now this. Later that week, she shook hands on the rope line in Charlotte, North Carolina, and over the hum of "Fight Song," a voice yelled out, "Secretary Clinton! Why did you say you were going to destroy coal jobs?" Trump loved contrasting Hillary's comment with his own vague promise to put "our steelworkers and our miners" back to work. His campaign released a video of his mushrooming crowds of mostly white men waving TRUMP DIGS COAL signs.

On the third day, she apologized. Asked if Hillary would talk to the press, one of The Guys said, "No, it's a self-immolation day."

Hillary later did what I called an Appalachian Apology Tour, a two-day bus tour through coal country during which an out-of-work coal worker, Bo Copley, confronted her as part of a roundtable discussion at a health center in "Bloody Mingo" County, West Virginia. "How could you say you are going to put a lot of coal miners out of jobs and then come in here and tell us how you're going to be our friend?" Copley said.

"I understand the anger and I understand the fear and I understand the disappointment that is being expressed. How could it not be given what's going on here?" Hillary told Bo. "Because of the misstatement that I made, which I apologized for when I saw how it was being used, I know that my chances [in Appalachia] are pretty difficult, to be honest."

I sat on the floor feet away and studied Hillary. She didn't win over Copley, a Republican who planned to vote Trump, but it had been one of her best moments. She'd shown empathy and that she had real plans to create "good paying clean-energy jobs." Mostly though, Hillary had shown up.

That same month, we were outside Tacoma, Washington. Tribal leaders bestowed on Hillary the Lushootseed Indian name meaning "Strong Woman" and wrapped a ceremonial Pendleton blanket over her emerald-green leather blazer, the one with the three-quarter sleeves and the fat buttons. The Travelers, the ten or so of us who stayed on to cover the Washington state caucuses, stood outside as a light mist fell on the Puyallup Indian reservation. Brown Loafers broke down the Hillary Gaffe Matrix.

"Number One: You get something wrong." [E.g., Hillary said, in 2008, that as first lady she'd arrived in Bosnia "under sniper fire." In reality, she'd been surrounded by children, Sheryl Crow, and Sinbad, who told the *Post* the "scariest" part of the trip had been deciding where to eat.]

"Number Two: You say something inartfully." [E.g., also in 2008, Hillary defended her choice to remain in the primary even though Obama appeared to be the nominee by saying, "We all remember Bobby Kennedy was assassinated in June in California."]

"Or, Number Three: You say something that you mean but you didn't intend to say." [E.g., months later, Hillary calling half of Trump's supporters a "basket of deplorables" would fit nicely into category three.]

The campaign argued Trump and the GOP had taken Hillary's coal comments so wildly out of context that it didn't even pass as a gaffe. "She was talking about creating clean-energy jobs! It was ridiculous," Brown Loafers said.

FLORIDA . . . WIN. NORTH Carolina . . . win. Ohio . . . win. Illinois . . . win. Missouri . . . win.

It was Tuesday, March 15, and Hillary—even with her gaffes—had cleaned the fuck up. For her victory speech, she chose West Palm Beach, down the road from where Trump delivered his own victory speech at Mar-a-Lago. "This is *another* Super Tuesday!" she said. Hillary had this incongruous urge to rhyme whenever she wanted to drive home a serious point. In her stump speech, it was "No bank is too big to fail, and no executive is too powerful to jail," and "Inversion? It's more like a *perversion.*" Was the country ready for a president who sounded like Dr. Seuss? "When we hear a candidate for president call for rounding up twelve million immigrants, banning all Muslims from entering the United States, when he embraces torture, that doesn't make him *STRONG*—it makes him *WRONG*," Hillary said in her West Palm Beach victory speech. She called on every American to fight the GOP front-runner's "bluster and bigotry."

In a campaign that had lacked electricity, that night in West Palm Beach was electric. Even Brooklyn thought she'd lose Ohio and Missouri—especially after the coal gaffe (or nongaffe gaffe depending on whom you asked). But she'd shown that scrappy underdog side Bill always talked about. On the Saturday night before St. Patrick's Day, she made an impromptu stop at the O'Donold's Irish Pub & Grill in desolate downtown Youngstown, a former steelworkers' metropolis that now has one of the highest unemployment rates in Ohio. I watched her down half a St. Patrick's Day pint of Guinness without taking a breath. "It's the best Guinness I've ever had! A Youngstown Guinness!" she said, throwing down what was left of the pint onto the wooden bar. A man in a DRINK LIKE A CHAMPION T-shirt leaned in for a selfie. "Born in the USA" blasted from a jukebox. "She knows where the Mahoning Valley is, that's for sure!" a woman yelled. "Something about Hillary tossing back a cold one makes days of miserable, tedious travel all worthwhile," Peter Nicholas said. I couldn't have agreed more.

IN MIAMI, HILLARY became a speedboat again. She shook hands with black and Latino hotel workers, a move I hadn't seen her do since the

Vegas casinos before the Nevada caucuses. I remembered that she did have a special connection to the working class—just not the white working class that everybody was talking about. A chubby-cheeked Afro-Cuban kitchen worker grasped one of Hillary's hands between his own and held her there for a second. He told me in Spanish, "If the bridges were all destroyed, I would swim to vote for her."

Rubio couldn't even beat Trump in his home state and dropped out. "America's in the middle of a real political storm, a real tsunami. We should've seen this coming," Little Marco said in Miami. Hillary's aides, almost all of whom descended on West Palm Beach, were elated—Rubio, whom they believed to be Hillary's toughest GOP rival, vanquished by a reality TV huckster. Hillary took a couple of days off in Chappaqua to regroup and let the reality of Trump sink in.

Bernie had the online donations to continue to wreak some havoc, but he was a zombie. Losing Ohio, Illinois, and Missouri meant he'd lost his post-Michigan luster as the white working-class whisperer. The Bernie Bros didn't want to accept this reality. They flooded my inbox with the literary flair of an unemployed liberal arts major whose prime interests were Holden Caulfield, the presidential campaign, and Internet porn.

"Patrick Healy and Amy Chozick are too busy sucking and fucking each other to spend a moment considering what Sanders is actually proposing . . ." They provided helpful feedback on our story about the Latino vote: "You're inept and biased and if you weren't so busy licking each other's assholes and dreaming of sucking Hillary's ass (well, you already are metaphorically) you might actually have time for journalism with integrity." The emails were usually signed the same way: "Go fuck yourself and I hope it hurts."

33

"Let Donald Be Donald"

Bill Clinton was the best friend Donald Trump always hoped to have. When a sex scandal loomed over the White House, Trump defended the president, saying he was a "terrific guy." His only criticism was of his choice. "It was Monica! I mean, terrible choice," Trump told the *Times*.

He'd urged the Clintons to move into one of his Trump-branded Manhattan condos when they left the White House. They opted for a five-bedroom colonial in Chappaqua instead. When a Waspy Westchester golf club hesitated to accept Clinton, Donald begged him to join the Trump National Golf Club, saying Clinton is "a great gentleman, a good golfer and a wonderful guy." (Both men are, in reality, mediocre golfers and cheats known for taking mulligans. Even Terry McAuliffe has pretended to enjoy eighteen holes and then thrown his clubs into the trunk of his car and sworn off ever playing golf with Clinton again.)

When Hillary won her Senate seat, Trump sent a congratulatory note (signed in black ink, and underlined twice, "Great Going!"). He extended invitation after invitation to host Bill and Hillary at Mar-a-Lago, but they preferred to vacation at Oscar de la Renta's place in Casa de Campo in the Dominican Republic.

When I asked Trump about this unrequited friendship, he reverted to marketing mode. "I don't recollect the Clintons ever staying at Mar-a-Lago in Palm Beach, Florida, which may be the most successful private club in the United States. If they did want to stay there, or stayed there, it only shows that they have good taste."

Less than a year before the campaign started, Bill and Trump golfed together at Trump National. Bill casually encouraged Trump to run, thinking his candidacy would roil the Republican field. For all his political foresight, Bill spent most of the primaries oblivious that this reality TV schmo could win the Republican nomination. He insisted that Rubio would pull through, even after his disastrous debate malfunction in New Hampshire. For one, Trump didn't really *do* politics. He was a notorious germophobe. Trump once saw Bill speak at a post-9/11 ceremony in New York and praised the former president, saying, "He shook hands with everybody out there. And some of these people had filthy hands."

When Rubio dropped out, the reality set in. Bill saw genius in Trump's economic populism and understood he was the perfect candidate for what Clinton called the "Instagram election"—an era when voters wanted only bite-size solutions. "Build a wall!" "Ban the Muslims!" "Make China pay!" Hillary didn't do bite-size.

"We live in a Snapchat-Twitter world," Clinton later unleashed on a crowd in North Carolina. "It's so much easier just to discredit people and call them names."

By late February, Bill went red in the face on almost daily conference calls trying to warn Brooklyn that Trump had a shrewd understanding of the angst that so many voters—*his* voters, the white working class whom Clinton brought back to the Democratic Party in 1992—were feeling.

He'd wanted Hillary to speak at Notre Dame, as he and Obama and Biden had all done. But Robby told him white Catholics weren't the demographic she needed to spend her time talking to.

Robby always listened patiently, respectfully. But he mostly saw in the former president a relic, a brilliant tactician of a bygone era. Behind his back, Robby did a Bill impersonation ("And let me tell you another thing about the white working class . . .") waving a finger in a Clintonian motion. Mook's mafia would laugh. Any Democratic operative under forty knew that those white voters were never coming back. The Hillary Coalition was built on suburban women, minorities, and the young. Trump had insulted so many voting blocs (women, Muslims, Mexicans, the disabled, Diet Coke drinkers, etc.) and provided such a veritable feast of offenses that Hillary, her top aides, and the DNC didn't think they needed to overthink the strategy.

Brooklyn had organized a book of oppo research on Rubio, Cruz, and Trump. The tome on Trump was the shortest, 157 pages, and mostly focused on his bankruptcies, his products being made in China, and some lewd comments he'd made on *The Apprentice*. ("It must be a pretty picture, you dropping to your knees.") They thought Trump would do the heavy lifting for them. We thought so, too. I cowrote a story about how suburban and independent women "who will play an outsize role in deciding the fall election"* would be so turned off by Trump insulting Heidi Cruz's looks that they'd run into Hillary's arms. Trump hated our premise, tweeting in response, "The media is so after me on women. Wow, this is a tough business. Nobody has more respect for women than Donald Trump!"

When the Heidi Cruz insult fest stretched into day three, Debbie Wasserman Schultz, the DNC chairwoman, told me, "I want Donald Trump to talk every single day for the rest of this election." Hillary summed up the strategy—or lack thereof—in simpler terms: "Let Donald be Donald."

I HADN'T WORKED out of the New York office in months. The first thing I noticed when I got back from Miami were the balloons.

In September, when I'd first put my "I'm driving long distances in Iowa . . ." out-of-office message on and prepared for life on the trail, the *Times Magazine* had dropped off a few extra balloons that they'd used in the photo shoot to illustrate a cover story about Trump by Mark Leibovich. The balloons were custom-made to look like Trump, with his sweep of mustardy hair across the top, a button mouth, and red power tie down the middle. They should've deflated months ago, but there was Trump's bulbous Mylar face, still buoyant and hovering over the *Times* politics pod.

---

* This was true. They just didn't end up voting for Hillary.

34

Stay Just a Little Bit Longer

MADISON, WISCONSIN, MARCH 2016

> Saturday, 3/19—HRC has no public events scheduled.
> Sunday, 3/20—HRC has no public events scheduled.

It went on like that for days, with a couple campaign stops in Arizona and California mixed in. Hillary figured she'd focus on fundraising and ride out the remainder of the late March and early April primary contests. Washington, Wisconsin, Alaska, Idaho, Wyoming—the white voters in those states were Feeling the Bern anyway. She didn't need them to win the nomination. Wisconsin, the only state with any real electoral value, hadn't voted for a Republican since 1984 when Ronald Reagan swept forty-nine states. The state's conservative talk-show host, Charlie Sykes, called Trump a "whiny, thin-skinned bully."

> Monday, 3/28—HRC will attend a fundraising event in Chicago, IL and then will deliver remarks on the Supreme Court in Madison, WI and hold an organizing event in Milwaukee, WI.
> Tuesday, 3/29—HRC will be campaigning in Milwaukee, La Crosse, and Green Bay, WI.

In keeping with her let-Donald-be-Donald mantra, Hillary decided to talk to the Travelers. On a stop at the Pearl Street Brewery, a chic industrial space in La Crosse, Wisconsin, with silver kegs, exposed pipes, and a foosball table, Hillary expressed her solidarity with the Breitbart

News reporter Michelle Fields who brought a battery charge against Trump's campaign manager Corey Lewandowski.

"I think that the reporter who brought the charge deserves a lot of credit for following through on the way she was physically handled at an event, and I think the charges being brought today certainly suggest that the authorities thought that her story was credible," Hillary said. Asked if she'd fire her own campaign manager if he'd done something like that—a scenario that seemed unlikely given Robby's admirable self-restraint—Hillary demurred, "I'm not going to comment on this particular case."

But she said that Trump's "negative and really mean-spirited language and actions" were to blame. "He is like a political arsonist, he has set some fires and then people have acted in ways that I think are deplorable." *De-plor-able, adjective, deserving strong condemnation; completely unacceptable; shockingly bad in quality*, the *New Oxford American Dictionary.*

Then Hillary held a pint under the tap and poured herself a Downtown Nut Brown ale. "I bet these reporters would like it!" she said and took a long swig. "Cheers! It's good."

The travel schedule was so blissfully light that I'd flown into Madison the previous morning and had time to squeeze in a spin class just off the University of Wisconsin campus before covering Hillary's speech about the Supreme Court opening left by Justice Antonin Scalia's death. She said that in Trump, Republicans got the hate-spewing candidate they deserved. "When you have a party dead set on demonizing the president, you may just end up with a candidate who says the president never legally was the president at all."

After that, Hillary went shopping. At a stop at the Anthology boutique, she browsed through beaded necklaces, magnets with vintage maps of Wisconsin, and other assorted tchotchkes. "Hello everybody, how are you?" she said to the Travelers as we squatted in the back of the kitsch-filled store. "We're determined to do some shopping. We've had a retail drought. I'm kind of looking, truly."

She settled on a red beaded necklace and a button that read KEEP CALM

AND VOTE HILLARY. She called out to Brown Loafers for some cash. As she headed for the door, Hillary looked back at the ten or so of us and said, "I love being in Wisconsin! I look forward to being here, meeting with people. We're just going to work hard."

> Wednesday, 3/30—HRC will be campaigning and attending fundraising events in New York City.
> Thursday, 3/31—HRC will be likely campaigning in New York and attending fundraising events in Massachusetts.
> Friday, 4/1—HRC will be attending a fundraising event in New Jersey.
> Saturday, 4/2—HRC has no events scheduled at this time.
> Sunday, 4/3—HRC has no events scheduled at this time.

That was Hillary's last trip to Wisconsin.

35

The Kids Are Alright

NEW YORK CITY, APRIL 2016
To say I didn't anticipate that Bernie Sanders would continue to be a pain in Hillary's ass until the very last contest in June, or that he'd ignite a burgeoning insurgency and roil the Democratic Party for years to come, would be putting it mildly. I initially brushed Bernie off with such casual nonchalance, such ill-informed, elite-media snobbery that I almost canceled our first one-on-one coffee because I didn't want to miss abs-and-back day at boot camp. It was April 2015 and Bernie's communications director, Michael Briggs, had reached out to see if I wanted to sit down with the Vermont senator at a Starbucks near Times Square. I'd said yes before I got word that Hillary would announce her candidacy in the next couple of days. Terrified that I would put on the campaign-trail twenty again and knowing that my mornings would soon turn into sciatica-and-gut days at a Holiday Inn Express somewhere, I wished I hadn't said yes. I dragged myself to the Starbucks inside the Sheraton on Seventh Avenue with the same mix of guilt and obligation as when my mom made me visit her emphysemic aunt Shirley. Shirley smoked her whole life and spent her golden years in a housedress hooked up to oxygen in a garden home at the Golden Manor Jewish Home for the Aged. We didn't have to bring a kugel to Bernie, but I did have to nod politely as he went through his talking points. I wrote in my notebook, "99 percent . . . new income in US . . . goes to top 1 percent . . . corrupt system . . . 'political revolution' . . ." I put that term, political revolution, in bitchy blue-ink quotes. I inched my chair away from the scent of black coffee and Quaker Oats that blew my way each time Bernie said "the

proliferation of millionaires and billionaires," with the stiff-lipped *B* and the long, breathy *ayy-ehrs* that would come to narrate the Democratic primary.

A couple of weeks later Bernie would announce his candidacy to a scant gathering of reporters outside Congress with such a blunt lack of fanfare that he prefaced the whole thing with "We don't have an endless amount of time . . ."

At the time, I couldn't envision how a socialist septuagenarian could almost defeat Hillary and in the process become a pop-culture icon. I'd picked up enough about economic policy covering the 2008 campaign through the financial crisis, including multiple interviews with Hillary, Warren Buffett, and Obama, to recognize that Bernie wasn't sure exactly how he'd make wealth more equitably distributed. I agreed with Hillary's assessment that Bernie didn't know anything. The more I pressed him for details on how he'd implement "a big bank breakup," as Larry David's *SNL* version of Bernie later put it, the brusquer and more irritated he became. He always returned to the same uncluttered, irrefutable point that the rich had become too rich and the American economy had gotten entirely out of whack. Toward the end of our chat, Bernie—as if he could hear my inner voice saying, *So when do we get to the part about Hillary?*—said "This is not about Hillary Clinton." No, he said if he ran his campaign, it would be about inequality, "about people working longer hours and earning less than they used to." It was John Edwards and the "Two Americas," minus the telegenic smile, the good hair, and the love child.

EXACTLY A YEAR after our first coffee, Bernie and I were back in New York. But this time I was one of hundreds of reporters who sat in the bone-dry, sunken fountain of Washington Square Park not far from where the beatniks rioted in 1961 and Allen Ginsberg read his poetry and the 2011 Occupy Wall Street protesters had set up tents. It was five days before the New York primary. There were people everywhere, twenty-seven thousand by one tally, "angel-headed hipsters burning for the ancient, heavenly connection to the starry dynamo in the machinery of night" . . . or, Bernie Sanders. They filled the streets of Greenwich Village weaving through police barricades and giving the area that had long since turned condo the throwback smell of some choice kind bud. The Bernie rally had the feel of Woodstock and the Triangle Shirtwaist Factory riots and an episode of *Girls* all rolled together in a springtime Feel-the-Bern bacchanal. I hadn't felt that kind of collective yearning, that massive gulped-down-the-Kool-Aid-and-asked-for-seconds crusade for a political candidate since I climbed onto the stage set up under St. Louis's Gateway Arch on a sunny afternoon in the fall of 2008 and looked out on an ocean of a hundred thousand people all chanting, "O-BA-MA, O-BA-MA."

A few weeks earlier in Portland, a delicate songbird had perched on Bernie's podium. "I think there may be some symbolism here," he said to the crowd of eleven thousand screaming fans. Hillary had been right to brace for defeats—#BirdieSanders went on to annihilate her in Washington and Wisconsin, Wyoming and Idaho, Hawaii and Alaska, collecting a paltry number of delegates but a hell of a lot of chutzpah. "Do not tell Secretary Clinton—she's getting a little nervous," Bernie said in Laramie, Wyoming. "But I believe we've got an excellent chance to win New York and a lot of delegates in that state."

His swagger filtered down to the Bros who believed the lamestream media and the Democratic Establishment were colluding to help Hillary. The Bros called me a "cunt playing the Cunt Card" who clearly needed to "take it like a man in the ass." The Bros told me that maybe if I hadn't spent entirely too much time "sucking Bill Clinton's dick," I'd appreciate Bernie's plan for free college. If I was delayed at an airport and bored, I'd write back, "You kiss your mother with that mouth?" Or on the rare occasion the Bros emailed me from a work account, I'd try to freak them out. "Are you speaking for yourself or is this the official position of Gen-Sales?" But mostly, I ignored them and told myself these were harmless white guys in their boxer shorts who couldn't find jobs or get laid and took some comfort in settling into their parents' basements to troll girl reporters. After all, I'd had my own toe-dip into online trolling (minus the oral sex references) when no one would hire me. Besides, the Bros weren't wrong. I didn't know fuck all about how they were feeling. I'd had no clue Bernie would become a geriatric Che, a white-haired white knight whose promise of ending "corporate greed and a rigged economy" would not only win over the kids but thrust them into open, angry revolt against Hillary.

AS FOR HILLARY, she was so done with her Esteemed Opponent that she could hardly stand to be on the same stage as Bernie at the Brooklyn debate. She didn't see any real difference between Bernie's peddling of empty promises to his hordes of sexist supporters and Trump's campaign, except that people seemed marginally better groomed at Trump rallies. One person who talked to Hillary about her views on Trump crowds vs. Bernie crowds broke it down to me as "at least white supremacists shaved."

One of Hillary's favorite moments in the primary—right up there with when she captured the nomination—came when Bernie nosedived in an editorial interview with the *Daily News*. He appeared unable to fully explain how he'd break up the banks. "Do you think that the Fed, now, has that authority?" the paper asked. "Well, I don't know if the Fed has that. But I think the administration can have it." After months of saying Wall Street executives should be jailed, Bernie wasn't sure if there were laws in place to bring about such indictments. "I believe that is the case. Do I have them in front of me now, legal statutes? No, I don't," he said. He even botched a question about how he rides the subway. "You get a token and you get in," Bernie said.

"Wrong."

Brooklyn blasted the transcript out to the press and donors ("MUST READ: Bernie Sanders' Enlightening Meeting with the Daily News Editorial Board"). They hung it on the walls. Hillary tried to rub in her New York credentials with a ride on the No. 4 train, even if it took her five swipes of her MetroCard to go through the turnstile at 161st Street in the Bronx. (HILLARY CLINTON'S METROCARD ADVENTURE: SWIPE. WINCE. REPEAT, the *Times* headline read.)

I WALKED ALONG the security perimeter of Washington Square Park. I asked people if they thought Bernie, who, as my profession liked to say, "lagged significantly in the delegate math"—as if a political revolution could be reduced to arithmetic—still had a chance to defeat Hillary. Aging hippies waved signs that said THE REVOLUTION IS HERE. A guy who looked exactly like a young Lou Reed held a poster with the words DEMOCRACY VS. OLIGARCHY, HUMANITY VS. GREED. Even if Hillary could've attracted the same all-you-can-eat buffet of random rallying cries, her corporate campaign regularly confiscated homemade signs. Brooklyn thought it best that the Everydays hold professionally produced signs that displayed the message du jour rather than something made with love and some finger paint and magic marker. In Phoenix, I watched a young Clinton staffer rip from the hands of a little girl an I ♥ HILLARY sign she'd drawn in crayon in art class that afternoon. They gave her a blue BREAKING DOWN BARRIERS sign with the campaign's H-arrow logo and hillaryclinton.com at the bottom.

I dove into the crowd like an anthropologist, eager to understand why young women, in particular, weren't With Her. But as I talked to so many students from NYU—and as their mouths moved and I followed up with "What's your major?" and "How do you spell Delilah?"—I was secretly seething with resentment. I'd wanted to attend NYU ever since our seventh-grade Hobby Middle School trip to Washington and New York.

I remember when our tour bus arrived in Greenwich Village, our faces pressed to the windows to gaze out at Christopher Street and Stonewall. I'd never seen men holding hands before, and I distinctly remember swooning over a woman with an emerald-green Mohawk and black leather vest with only a black lace bra underneath. That's when Brandy said, loud enough so that I could hear her several rows in front, "Amy's so weird, I bet she wants to live here." She was right. From that point on,

I was determined to go to NYU and live in a rundown dorm on Mac-Dougal Street and pierce my nose and wear a pleather skirt to the Tunnel every Thursday night.

For the next year of middle school and four years of high school, I wore an NYU T-shirt at least twice a week and acted as if going there was a done deal. I'd say, "I'll be in New York by then . . ." and "I'll have to live in Greenwich Village to be close to my classes . . ." But I'd never actually talked to my parents about this plan until my junior year.

One evening I tiptoed into my parents' bedroom in my tattered NYU shirt holding a folder, with NYU's white-and-purple torch, that laid out the various tuition and financing options. My dad lay on the bed as he usually did late into weeknights, legal papers spread out over his lap and his brown leather briefcase propped open like a bear trap beside him. My mom sat on the floor grading papers. I'd rehearsed all of it in my head for so long. "I could do a work-study program . . ." I started. "I've got all this experience hostessing already, and they have this program where . . ." NYU wasn't up for discussion. "We've talked about this," my dad said. "There's a perfectly good state school an hour away in Austin." In trying to prove how adult I was, how ready I was to live in New York, I threw an absolute balls-out, arms-flailing temper tantrum. "YOU DON'T KNOW ANYTHING," I sobbed. "NYU isn't a state school. It's PRIVATE and *PRESTIGIOUS*." At this point, my sister Stef, who'd been on my side about fleeing Texas (she'd dreamed of Georgetown before ending up at UT), heard my fit and turned on me. "You seriously think Mom and Dad are going to pay twenty-five thousand dollars a year so you can go be a druggie in New York?"

I didn't even mail in my application. I spent the fifty-dollar application fee on tickets to the Primus concert and a bag full of hallucinogenic mushrooms that my neighbor Travis grew on cow patties. I still got my nose pierced, a silver loop too big for my face that sat in a dollop of pink pus on my left nostril. This accessory lasted exactly two months until my mom told me they'd stop paying my UT tuition if I didn't "take that crap out of your nose this second."

I looked at these twiggy, unshaven girls living in the West Village on their parents' backs. They wore gray wigs with peach latex designed to look like Bernie's super-hip receding hairline. They had tight-fitting T-shirts with Bernie's black glasses and slogans like BERNIN' WITH PASSION and BERNIE IS BAE and BROOKLYN FOR BERNIE. My envy began to fade. I'd been a brat. My dad had been right all along. The perfectly good state school an hour away—and $4,000 a semester—had been exactly what I needed. But I couldn't see it then. I must have seemed to him like one of the unwashed masses who couldn't see that Hillary was the obvious, practical choice.

Tim Robbins stood onstage under the miniature, lit-up Arc de Triomphe at Washington Square Park. "We are supporting a candidate that has taken principled positions when others have compromised," he said. "What a radical concept: a politician that has a moral bottom line."

Democrats were facing a general election against Trump, and Bernie's campaign was still saying *Hillary* was the one who lacked a moral bottom line? She was a Methodist minister compared to Trump. But you'd never know talking to the crowd that night. The dozens of Bernie lovers I interviewed told me they wanted to read the transcripts of Hillary's Wall Street speeches. They hadn't forgiven her for voting for the Iraq War. They bemoaned her billionaire donors and her pussyfooting around whether she supports a fifteen-dollar minimum wage ("I think setting the goal to get to twelve dollars is the way to go, encouraging others to get to fifteen," she said in the Brooklyn debate). They said she'd shamed Monica Lewinsky and that she'd been "bought and paid for" by corporations. Everything Trump would throw at Hillary and then some, hurled from the glossy lips of college girls that night.

THE DAY BEFORE the New York primary was a twenty-seven-hour whirlwind trailing Hillary as she posed for photos with Latino workers at the Hi-Tek Car Wash & Lube (slogan: "Lube it or Lose It") in Elmhurst ("I've campaigned at lots of gas stations, some of which had car washes, but this is special!"), sipped tapioca bubble tea in Chinatown in Flushing

("Hm, I've never had squishy tea before"), and, at an East Village arti-sanal ice-cream shop owned by an ex-con, broke her rule of never eating in front of the press and devoured a milk chocolate and waffle ice-cream concoction named the Victory. It ended that night with the Irish. The Travelers filed into the courtyard of the Fitzpatrick Hotel near Grand Central, a little gem, owned by the donor and eligible bachelor John Fitzpatrick, that my Irish family (and no one else I knew) consider the height of New York luxury. Bill Clinton, in a green tie, and an Irish fid-dle player warmed up the crowd. "Too long a sacrifice can make a stone of the heart. Hillary has spent an entire lifetime taking the stones out of the heart by making good things happen." I'd heard Bill butcher lines of Yeats for years, but that one seemed particularly creative.

After they learned that the *Times'* Hillary reporter is "married to a Meath man" (as in County Meath), New York's Irish power brokers adopted me as one of their own. I had a lot of sleazy sources whom I'd never spend time with if they weren't potentially useful (see Hands Across America). But the Irish were like family. I loved their dark sense of humor and morose worldview. In the midst of the New York primary, I received anti-Semitic mail, including a brochure that showed a stereo-typical Jew with a black hat and exaggerated, elongated nose, rising above the earth and engulfed in voracious flames, over the words "EVIL JEWS IN CHARGE!" An Irish friend took one look at it and shrugged, "Eh, at least you're in charge."

I stayed around to drink a Magners, the Irish cider that I'd learned to order in lieu of Guinness. Strings of translucent lights hung overhead, and the fiddler played the reedy chords of "Rocky Road to Dublin." I asked Bill how he'd first bonded with the Irish. This led to a ten-minute walk down memory lane about how the Irish had helped a redneck gov-ernor from Arkansas win the New York primary in 1992, and how in turn he'd approved Gerry Adams's visa and helped end the Troubles. "You didn't used to care about this when you went to New York, you just wanted the Irish to vote for you . . ." he said.

I got home after midnight. I'd only had one pint but felt drunk. We had to be back on the van at 5:45 a.m. to watch Hillary vote in Chappaqua.

That morning a light drizzle fell outside Douglas Grafflin Elementary School where Hillary's well-heeled neighbors stood under golf umbrellas and held handmade signs and chanted "Hillary!" and "We love you!" Inside the basketball gym, Hillary leaned over to cast a ballot for herself. The Travelers barreled around her. We yelled our evergreen question, "Secretary! How are you feeling about tonight?" Hillary snapped, "Guys, it's a *private* ballot," and signaled to Brown Loafers. "Can we get the press out of here, *please*?"

By the time Hillary came outside, the rain had stopped. The clouds had parted and the sun beat down. Again, that question, "How are you feeling about tonight?" Hillary looked over her shoulder at us. "I love New York," she said.

She bulldozed Bernie by sixteen points and came onstage at the Sheraton that night to twenty-five hundred screaming New Yorkers and the slow beat of "Empire State of Mind" filled the ballroom—"Concrete jungle where dreams are made of / There's nothing you can't do."

I'd already filed my story and lifted myself over the barricade set up around the press area to watch. I'd asked for an embargoed draft of Hillary's remarks and a source-friend in Brooklyn instead texted me a meme of Sally Field accepting her second Oscar, arms hoisted overhead and the words, "You like me! You really like me!"

36

Writing Herstory

I often want to cry. That is the only advantage women have over
men—at least they can cry.
　　　—Jean Rhys

BROOKLYN, JUNE 2016
At 8:21 p.m. ET on Monday, June 6, 2016, Hillary made history in the
most Hillary way possible. The Associated Press flashed across the wire:
BREAKING: @AP finds Clinton reached the number of delegates
needed to clinch the Democratic nomination for president. Hillary
had done something no woman in the 228-year history of the republic
had ever come close to doing. Thing was, California hadn't voted yet.

　　She'd planned to do the whole history-making hoorah the following
night in Brooklyn, under the glass-domed ceiling of the Brooklyn Navy
Yard, eight years to the day after she dropped out of the 2008 primary.
But the AP called a bunch of uncommitted superdelegates, those pesky
party elders who got to vote for whatever candidate they wanted regardless of the will of the voters. That got Hillary to the magic 2,383 number
of delegates needed to become the nominee. She didn't want to win that
way. The Bernie camp was already pushing the idea that Hillary would
only win because of the party's funky, undemocratic electoral system.
She'd win because of the ESTABLISHMENT.

　　Hillary stood backstage at the Greek Theatre in Los Angeles when
she saw the headline. She waited a while to go onstage as John Legend

introduced her and downplayed the news. "I don't care what the AP says about who won the nomination," he said. "We need everybody to vote tomorrow." She didn't care about the other five states voting on Tuesday, but she couldn't lose California, couldn't even be close, or everyone would talk about her weaknesses, and we'd all write that even though she made history she "hobbled to her party's nomination."

Earlier that afternoon I was in the New York office watching a gaggle on TV. Hillary had been shaking hands at a senior center in Compton when the Travelers gathered around to ask about the heft of what she'd accomplished. She responded with her usual stoic restraint. "I am really just so focused on all of the states that are voting tomorrow. That is my singular focus because I know that there's a lot of work still going on." Pressed again, Hillary didn't waver. "I'm going to wait until everyone has voted. Tomorrow night we will have a chance to talk more about this." I thought back to one of the low points in the fourteen-month campaign, when a woman at a town hall in Vegas asked Hillary for a hug, and she'd replied, "Sit down right there. When I finish my Q&A, I will give you that hug, I promise." *Sit right there, and when I kick Bernie's ass in California, I'll acknowledge the historic moment I've spent decades dreaming of and that generations of women have hoped for, I promise.*

When the AP headline moved across the wires, I was on the F train and specifically in the black hole between West Fourth Street and Broadway-Lafayette, the only stretch on my commute when I lost service entirely.

8:27 p.m. ET. "Next stop, Delancey-Essex," the conductor muttered. That's when I saw the tweets, saw all the missed calls and voice mails, scrolled through the frantic emails flooding in. I felt the stab in the gut when I read a note from Carolyn: "Maggie and Nick did a quick cut."

For eight years, ever since Hillary declared under the atrium of the National Building Museum in Washington that 18 million cracks had been made in "that highest hardest glass ceiling," I'd imagined the night she'd finally become the nominee and thought about what my story would say. Now, after two presidential campaigns and five months of primaries and caucuses, it had finally happened, and I'd been in the bowels of the Metropolitan Transportation Authority as a couple of colleagues who happened to be in the newsroom wrote the first draft.

In the approximately six minutes I'd been underground, a heated newsroom-wide debate had erupted about how to handle the AP's call. Everyone had an opinion . . .

"Nobody asked me but isn't this a bit of an AP news gimmick?"

"FWIW, this is also how the AP got to the twelve hundred and thirty-seven number for Trump—calling superdelegates and getting them to commit on the record."

"It's like the AP is 'reporting' her way into the nomination, as opposed to waiting for her to win it tomorrow. At this stage after months of primaries and caucuses and a big day tomorrow, I wouldn't go big with this."

"They did the same damn thing in 2008."

"You could write a story attributed to the AP but hedging (if you're not ready to buy the whole hog) or you could declare it over and run the story you intended to run tomorrow."

"I can imagine some of the more devoted Sanders fans have some strong feelings about the timing of this."

I was a block away from our apartment, passing the Chinese bodega with the handmade WE SELL BEER sign and scrolling through the emails to try to figure out if anyone had come up with a plan. I landed on Carolyn's definitive "Really need everything." message. I picked up into a jog. My colleagues had saved my ass and gotten a couple of quick graphs onto the home page as a placeholder until we could get a longer version online and in the next day's paper. My back wasn't spasming the way it usually did when I ran under the weight of my laptop, chargers, a change of shoes, and a few bottles of water. When I stopped to look for my keys, I realized why my backpack felt light. I'd left my laptop at the office. By the time Bobby heard me rummaging around for my keys and opened the front door, I was hyperventilating.

For weeks, I'd started to pour everything I knew about Hillary's life, all her contradictions and what her career meant for women and our rapidly changing expectations of ourselves, into a tight twelve-hundred-word story that would run when she captured the nomination and which was partially written on a laptop that was half an hour away in Times Square. I tried to use Bobby's computer—a brick of an old HP that we named Big Eddy. But after five minutes trying to browse Internet Explorer, Big Eddy overheated and crashed. I called the office. Nick Corasaniti, a Trump reporter who appreciates cheap eats and surfing in that order, was about to leave for the night. I lured him to the Lower East Side with my laptop and the promise that he could also pick up a poke bowl at Dimes, the California-cool eatery down the street. I couldn't stop working long enough to run downstairs. Bobby went down for the laptop, sprinted upstairs, and stretched my MacBook Air toward me like a relay race runner passing the baton.

Pat had been at the gym when the news broke. By then, we'd teamed up on dozens of stories about the Democratic primary. But Pat covered the Republican primary, too. Pat covered everything. Trump favored him, calling him "smart Irish." Pat could churn out a perfectly crafted front-page *Times* story during a commercial break with sixteen colleagues, five editors, and a cranky copy editor shouting unsolicited advice at him. I needed his help with this one. He wiped off his sweat and headed to the newsroom so we could team up on the story.

We filed the story in two-hundred-word chunks, as we wrote it, so that editors could piece it all together in real time. The copy desk came up with the careful headline CLINTON REACHES HISTORIC MARK, A.P. SAYS. The "I wouldn't go big with this" contingent lost. The story ran across six columns on the front page.

"THANKS TO YOU, we've reached a milestone—the first time in our nation's history that a woman will be a major party's nominee for the president of the United States!" Hillary said twenty-four hours later in Brooklyn.

In 2008, Mark Penn had advised she run as a man, and I'd raced out of an auditorium in Salem, New Hampshire, to try to talk to the man who'd waved a yellow sign and interrupted Hillary with chants of "Iron my shirt!"

In her potpourri of a 2016 announcement speech a year earlier on Roosevelt Island, even the feminist lines were sort of about men. Hillary said she dreamed of "an America where a father can tell his daughter: Yes, you can be anything you want to be, even president of the United States."

Bernie's supporters, Republicans, and garden-variety Hillary haters always told me it wasn't about gender. They'd vote for a woman, just not *THAT woman*. Hillary walked onstage in Brooklyn wearing white for the suffragettes. I wanted to scream at every critic that thirty years of sexist attacks had turned her into *that woman*. That sooner or later, the higher we climb, the harder we work, we all become *that woman*.

Hillary stretched out her wingspan wide enough to embrace the entire screaming crowd, all the little girls hoisted on their fathers' shoulders. Okay, so most of their fathers were Wall Street donors, but she spoke in language of social activism, always *we* instead of *I*, as in "the history *we've* made here." She took the stage after a video showed the women of Seneca Falls, Shirley Chisholm, Gloria Steinem, and Hillary's own "women's rights are human rights" declaration in Beijing, putting the evening into the context of decades of struggles in the women's movement. I'd sent an early glance of the introductory video around the newsroom. A male editor replied, "Where are all the men?" The *Drudge Report* headline read I Am Woman Hear Me Roar.

Hillary devoted the victory to her mother, Dorothy, born June 4, 1918, the same day Congress passed the Nineteenth Amendment that would grant women the right to vote. "I wish she could see her daughter become the Democratic nominee for president of the United States," she said.

She wasn't a blubbering Lifetime movie, but at this point, Hillary was clearly fighting back tears. In 2008, I'd been a couple of feet away from Hillary at diner in Portsmouth, New Hampshire, when the waterworks burst. Jamie had stepped outside for a second and missed it. The Guys called panicked asking her what the hell happened. The teary moment became an earth-shattering inflection point debated by political pundits and historians for years to come. I considered it progress that eight years later, Hillary cried all the time and no one really noticed.

There was the time backstage in Manchester when a part-time librarian told Hillary his eighty-four-year-old mother had Alzheimer's and he couldn't afford her care, "so I take her to work with me." Her eyes welled. "Oh my gosh. I'm sorry I didn't mean to . . ." And there was the town hall in Keota, Iowa, when ten-year-old Hannah Tandy asked Hillary what she would do about bullying. "Can you tell me a little bit more about why that's on your mind?" Hillary asked. "I have asthma and occasionally I hear people talking behind my back," Hannah said. A teary Hillary pulled Hannah in under one arm. Brooklyn promptly turned the interaction into an anti-Trump ad.

I'd been more sentimental than usual, too. When Hillary said her mother "taught me never to back down from a bully, which, it turns out, was pretty good advice," my thoughts drifted to the Bernie Bros. I'd spent most of my day the way the other women covering Hillary had— fending off death threats from the Bros who thought we prematurely called the race in Hillary's favor. The colorful strings of expletives I'd grown used to, even amused by, became more violent. Several Bros called my number at the *Times* to tell me the revolution was coming for me. One of them left me a voice mail about how he knew my type of "rich bitch." Others seemed to read from a script that went something like "You lying, Hillary-loving cunt. We will hunt you down in the fucking streets."

I needed to bask in some girl power or at least commiserate with my fellow Hillary-loving cunts. Before Hillary took the stage, on an esplanade of the Brooklyn Navy Yard overlooking the Brooklyn Bridge, my closest female friends among the Travelers—Ruby Cramer, Annie Karni, Jennifer Epstein—and I posed for a photo. We are in our typical position, guzzling down iced coffee and cradling our laptops as if they were newborns. We made it through the primaries. We were there to write "Herstory," as JenEps called it, and for about fifteen minutes, it felt damn good.

I was back to reality when I ended up watching Hillary's speech through the isosceles triangle of a photographer's denim thighs after he'd parked his schlubby ass right in front of me on the press riser just as she took the stage. Herstory, as witnessed through the filter of a male crotch.

Making matters worse, I looked over at the second press riser, even closer to the stage, and saw a shih tzu that appeared to have Bell's palsy, his pink tongue falling out one side of his furry, lopsided face, sitting on a cozy pillow with a prime view. After the speech, I climbed off the riser and raced over to Hired Gun Guy to rip into him that a dog had been assigned a better position on the press riser than the *Times*.

"That fucking dog and his little doggie bed had a prime view," I'd said.

"You're kidding me, right?" Hired Gun replied. "That's Marnie the Dog. She has like two million followers on Instagram. Sorry, but the shih tzu has more reach than the *Times and* the AP."

THE NEXT DAY, another story I'd written in my head ever since Hillary's 2008 concession speech ran on the front page under the headline HILLARY CLINTON'S LONG, GRUELING QUEST. To me, the single moment that encapsulated Hillary's path to the Democratic nomination wasn't "her sun-splashed campaign kickoff in New York," or her speeches "celebrating hard-fought primary victories," but the unscripted instant during her eight hours of testimony to the Republican-led Benghazi committee when "a blasé Mrs. Clinton coolly brushed from her shoulder a speck of lint, dirt—or perhaps nothing at all."

The next day Hillary agreed to give the print reporters interviews, something we'd been collectively requesting for months. I had twelve minutes on the phone that I stretched into fourteen minutes and thirty-five seconds, only because I turned the conversation to our summer reading lists. Hillary said she planned to hit the Chappaqua bookstore. "I love to wander around bookstores and see what strikes my fancy and hear what the people who work there recommend," she said.

I asked her what the last book she read was, and she recommended

Diana Nyad's memoir, *Find a Way*, about her nearly 111-mile swim at age sixty-four from Cuba to Florida.

"It is something that when you're facing big challenges in your life you can think about Diana Nyad getting attacked by the lethal sting of box jellyfishes and nearly anything else seems doable in comparison," Hillary said.

"Is Donald Trump the box jellyfish in this scenario?" I asked.

Hillary let out one of those guttural Hillary belly laughs. "I don't know about that."

37

Who Let the Dog Out?

A TARMAC IN PHOENIX, JUNE 2016

In the span of a couple of months, Bill Clinton's conversations on private-plane tarmacs went like this: He chatted with Orrin G. Hatch in Louis-ville after speaking at Muhammad Ali's funeral. He ran into Paul D. Ryan and gave the Republican speaker a polite piece of his mind. He said a quick hello to Arnold Schwarzenegger and, in Mobile, Alabama, he even shook hands with Ted Cruz, despite the Texas Republican having recently called for Hillary's imprisonment.

But in late June, with the Democratic primaries over and everyone saying the fall election was Hillary's to lose, Bill made twenty minutes of tarmac small talk that altered the course of the race.

In the midst of a ten-city fundraising swing, he stepped off the scorch-ing asphalt at the Phoenix Sky Harbor International Airport and onto the airplane of the attorney general whose Justice Department was inves-tigating his wife's handling of emails at the State Department. It took a while for anyone in Brooklyn to learn about the tarmac chat with Loretta Lynch, and even then they didn't initially see it as much of a problem.

Josh Schwerin, the press aide who had accompanied Clinton on the trip, didn't think to tell Brooklyn. Josh's status as the youngest of the Schwerin brothers, after Ben, who worked in the Clinton White House, and Dan, who had interned in Hillary's Senate office until he worked his way up to the chief speechwriter position, protected him. But for a few days, when Lynch distanced herself and said she'd accept the FBI direc-tor James Comey's decision in the email investigation, the well-meaning kid thought for sure he'd be fired.

I sympathized with Josh. In keeping with my inability to gauge Clinton scandals, I didn't initially realize that a little tarmac small talk would become a major story either. When I first ran it by my editors, they told me I could skip it.

I don't know why Clinton did it. Lynch later said Clinton came over mostly to say hello and "talk about his grandchildren and his travels and things like that" and that "no discussions were held into any cases or things like that." The worst Democrats would say about the meeting on the record was that it had been "irresponsible."

The Guys returned to the lovable St. Bernard defense. *He just wanted to say hello to an old friend and talk about the grandkids.* There were theories that Clinton subconsciously didn't want Hillary to be president, but I didn't believe those, especially not when it meant his surname would forever be associated with giving the world Donald Trump.

FOBs told me he probably was trying to influence Lynch, not directly, but just a mild charm offensive that he thought would help Hillary.

"OTR—" read a text message from a source who knew Clinton better than most, shortly after news of the meeting broke. "President Clinton is 'irresponsible' like a fox."

38

"Man, Y'all Are Jittery"

Fear can be conquered. Anxiety must be endured.
—Max Brooks, *Minecraft: The Island*

NORTH CAROLINA → FLORIDA, JULY 2016

We all stood there in our own worlds—me; Michael Shear, a White House correspondent; and Matt Flegenheimer (aka Fleg), a wunderkind who'd been on the Ted Cruz beat and would now help out with Hillary coverage. We'd all flown into Charlotte, North Carolina, to cover Hillary's first joint rally with Obama. But as soon as we landed, we learned that the rally wasn't the news, not even close.

I'd spent the Fourth of July weekend focused on two activities. Continuing my habit of overcorrecting and going full Martha Stewart when I was home, I baked Bobby a flag cake and decorated it with white icing and rows of strawberries and blueberries. And I'd written about Hillary, accompanied by five of her lawyers, sitting down for three and a half hours of questions from eight FBI and Department of Justice officials. In any other election year, the sight of the leading candidate for the presidency and her phalanx of lawyers strolling into the J. Edgar Hoover Building would have shattered the political universe. But this was 2016.

By late morning on July 5 when our American Airlines flight landed at Charlotte Douglas International Airport, the FBI director, James Comey, had given a press conference in which he recommended no criminal charges be brought against Hillary's use of private email. But (and with

Hillary, there was always a but) he also said she and her State Department aides had been "extremely careless" in handling classified information.

We stood at the Hertz rental car counter alternating between checking Twitter and throaty gulps of iced coffee. We shook with the edgy irritation that only comes from flying to a place thinking you'll write the big story and then realizing once you've landed that you've missed the story entirely. Adding to this agitated stew was the dynamic with my newly installed sidekicks on the Hillary beat: Fleg, with the slumped posture, sense of humor, and prose style of a *Times* reporter several decades older than his twenty-seven years; and Tom Kaplan, a menschy boy genius with too many Ivy League degrees to be sending me feeds from a town hall in Akron. I adored both like the little brothers I always wanted, and the travel and daily story demands were too much for any one reporter. But I'd never shared the Hillary beat before. I remembered my days as a starting point guard in Texas prone to offensive fouls. *Mine, mine, mine.*

We tried to stream highlights of the Comey press conference, while all talking into our earpieces at once. "No, the story needs to run *tonight* . . ." and "Just landed. Need anything from me on the Comey presser?" and "Does this mean we're bumped off the front?" The Obama endorsement became such a nonstory that our editors asked us to "chunk it up." In *Times*-speak, Chunky Journalism means the listicles of photos with snappy captions, tailor-made for millennials to scroll through on iPhones. I'm not opposed to the idea, but I hated the name. I didn't want to be called chunky and neither did my journalism. When I raised this concern, a politics editor proposed Tapas Journalism instead. "Who doesn't like tapas?"

That's when we realized the woman behind the Hertz counter was looking at us up and down, her eyebrows peaked. She shook her head and, as she handed over the keys, said in a slow Southern timbre, "Man, y'all are jitt-ehr-eeeeeeeee."

Two weeks later, on a hotel patio overlooking a man-made lake in Orlando, I reached peak jittery.

Everyone was in Cleveland to cover the Republican National Convention, but I'd flown into Orlando the night before to meet up with

Hillary. It takes a lot to make me envious of a Red Roof Inn on the out-
skirts of Cleveland, but being left behind for the sole purpose of trying to
break who Hillary would choose as her veep was worse. Trying to scoop
the veep was one of those competitive sports in campaign reporting that
made me wish I were still wandering around Shibuya writing stories
about oxygen-infused water and the latest hot-tub karaoke craze. The
history books hardly took note of running mates, never mind the hack
reporters who spent months trying to break the story. Get it right, and
you were rewarded with the warm bath of three minutes tops of praise
on Twitter from a tiny clique of political reporters. But get it wrong, and
you'd live in DEWEY DEFEATS TRUMAN infamy—which very nearly happened to me eight years earlier.

I was sitting on the toilet at the InterContinental Hotel on Chicago's
Magnificent Mile with my BlackBerry in hand waiting for the late-night
text from the Obama campaign to announce his vice-presidential pick.
My *Journal* editor, whose steady neurosis and Harvard economics degree
made him both brilliant and unsuited to steer political coverage, sent
out a news alert that went something like WSJ BREAKING NEWS:
SENATOR BARACK OBAMA CHOOSES VIRGINIA GOVERNOR TIM KAINE AS
HIS RUNNING MATE.

I bolted up as if an earthquake shook the ground beneath me and,
pants around my ankles, called the desk. "Take that down immediately!
I never said it was Kaine! TAKEITDOWNNOW!"

I'd escaped calamity that time, but there would be no way as a *Times*
reporter in the Twitter era that the Internet would be so forgiving again.
I'd fallen off Carolyn's radar as she focused on the GOP side. There was
only one way to regain the twinkle Carolyn bestowed on the reporters
who brought to her the biggest, the yummiest, the hardest-to-get scoops:
I needed to break the veep news. I had to make Mamma happy.

I'D SPENT A week practically alone in the newsroom chasing every wacko
lead. I read two books by James G. Stavridis, a retired four-star navy ad-
miral, before a senior aide told me he gave new meaning to the words short

list. "Have you seen James Stavridis?" (He stood a couple of inches shorter than Hillary, without her kitten heels.) I recruited a body-language expert to study Hillary's demeanor, including the brush-off she gave Cory Booker after he spent thirteen minutes sucking up to her at a rally. This included quoting Maya Angelou, Abraham Lincoln, and Jon Bon Jovi. "I hate to contradict Bon Jovi," Booker said, "but dear God, Hillary Clinton, you give love a good name." Hillary patted him on the back, a gesture that reminded me of the time Matt Paul, her campaign's former Iowa state director, had tripped as he shuffled alongside the rope line next to her. "That's Matt Paul," Hillary said. "He's from Iowa and he's doing the best he can." *That's Cory. He's from New Jersey and he's doing the best he can.*

I'd flown to Orlando the night before and watched Trump's I-alone-can-fix-it acceptance speech in my heavily air-conditioned hotel room while texting with one of my most credible sources. I'll call him Sean. Sean assured me Kaine was the guy. ("They figure if she wins Virginia, Trump has no path.") Jen Palmieri, intent on avoiding leaks, could smell on me the fragrant stew of desperation and self-doubt. "Your veep source," the subject of one email read, "is either terrible or deliberately messing with you."

I held an eyelash between my finger and thumb and blew into the humid July air. "I wish this day were over and the veep news was behind me."

A DRAFT OF my "Hillary Chooses Tim Kaine as Running Mate" story was edited and ready to publish. I told my editors to hold off and texted Sean again.

"What's the latest? You still hearing Kaine?" I patiently gave him two full minutes to reply and texted again. "Are you there? Are you pissed at me?"

The bouncing dots appeared . . . "Don't worry, nothing changed overnight. She's not going to cross Bill, never has, and has McAuliffe support, too."

The Travelers were even more high-strung than usual. With Cleveland behind them, all our editors had turned to Hillary. They called us all nonstop under the laughable assumption that traveling with the campaign must mean we knew what was happening in the campaign. Hillary held a roundtable discussion with community leaders to talk about the aftermath of the Pulse nightclub massacre a month earlier. After that, she went to Pulse to lay a dozen white roses at a makeshift vigil for the forty-nine victims. I watched Hillary walk past Catholic prayer candles in glass jars, rainbow flags, bunches of fresh sunflowers, and a collage of faces so young they could've been high school yearbook pictures. Whom she picked as her veep had never felt so beside the point.

The day dragged on. The "Hillary Picks Kaine" story sat unpublished. The text messages flooded in. We were at the Florida State Fairgrounds in Tampa when one of the possible veep picks texted to tell me Podesta had called him to tell him it wasn't him. Sean texted again: "Any second. Go for it." A press aide yelled, "LOADING!" and she led the Travelers through a pitch-black warehouse adjacent to the fairgrounds. I could hear the stampede of the Travelers, but couldn't see anything. My iPhone lit up. It was my colleague Jonathan Martin, who knew Virginia politics better than anyone. "KAINE." A second later, "GO WITH IT!" I called the desk and told them to publish. The rush of adrenaline, the darkness, the sense of dread if we didn't get it right, all made me feel as if I were wading through a haunted house. Roughly eight minutes later, Brooklyn blasted out the official announcement. The text to supporters read, "I'm thrilled to tell you this first: I've chosen Sen. Tim Kaine as my running mate. Welcome him to our team, Amy."

Hired Gun Guy tried to smoke out our source. "Based on your sourcing, should I assume someone on the campaign confirmed it ahead of the text?" he emailed. I took more satisfaction than I should have in sending back an emoji smiley face with his lips zipped.

Brooklyn settled on another theory to explain how the news had not only leaked but been attributed to a source inside the campaign. They thought I'd pretended to use the bathroom at the Tampa fairgrounds and eavesdropped on Jen and Huma, who stood behind a nearby curtain and loudly discussed the logistics of Hillary and Kaine's meeting later that night in Miami. I wish I'd thought of that.

AFTER WEEKS OF misleading us and telling us our sources were full of shit, The Guys climbed onto the press bus to make sure we all wrote the same dramatic, highly favorable tick-tock. Brown Loafers delivered a readout that went like this: Podesta had visited Hillary in Chappaqua, after the New York primary, with Duane Reade bags full of binders with background on potential vice-presidential candidates. He told us how Bill, Hillary, and Chelsea bonded with the Kaine family over lunch, and how aides had slipped out of Brooklyn via a freight elevator to usher Kaine away from a Newport fund-raiser and fly him incognito to Miami. We all wrote down "April" and "Duane Reade bags" and "Newport fund-raiser." The headlines came out exactly as planned: BEHIND THE CHOICE: HOW HILLARY SELECTED HER RUNNING MATE and CAR CHASES AND SECRET GETAWAYS: TIM KAINE'S WILD 78 HOURS.

THE GUYS DIDN'T mention that Podesta and Cheryl started the vetting process way before the New York primary. In mid-March, Podesta passed on a list of possible veeps arranged into "food groups." This balanced diet included Latinos (Tom Perez, Julian Castro), women (Elizabeth Warren, Kirsten Gillibrand), blacks (Cory Booker, Eric Holder), white male politicians (Tim Kaine, Terry McAuliffe), military men (John Allen, Mike Mullen), and assorted billionaires and businessmen (Bill Gates, Mike Bloomberg, Tim Cook). Thirty-ninth, and last, on the list in a food group all his own was Bernie Sanders.

THE PRESS CHARTER to fly us to Miami that night hadn't even started to taxi when I pulled down my tray table and popped open a Pabst Blue Ribbon. I closed my eyes and listened to John Coltrane and felt my restlessness recede into the tenor saxophone of *A Love Supreme*. With the veep pick behind us, the end was in sight. I could do this.

In three months, the campaign would be over. The Steel Cage Match would disassemble. I could put a close to my years of only covering Hillary in the defensive crouch of a presidential campaign. I zoned out the window at the dark swampy air, wondering what Hillary would be like to cover as president. Sure, we had a convention and a general election to get through, but how bad could that be?

And right when I expected to feel the skip of the wheels as they lifted off the tarmac for Miami, we were told there was a "crew rest" issue and the plane couldn't fly. The pilot and flight attendants started to deplane. Brown Loafers—who on any other night would've flown on Hillary's jet, which was well on its way to Miami—called Brooklyn to figure out how to get us out of Tampa. The prognosis wasn't good. We had two options: We could track down a bus and drive the 280 miles to Miami, or we could spend the night at an airport hotel and fly out at six the following morning. Brown Loafers burst from his seat in disbelief. "Stuck in the great coastal shit hole of Tampa Bay!" Bill Nelson, the Florida senator and former astronaut who flew on the mission right before the *Challenger* exploded, had been at Hillary's rally and hitched a flight to Miami on the press plane. Hearing this debate unfold, somebody suggested that Nelson fly the plane. This led to a split among the Travelers.

"Is that a good idea?"

"Yeah, I don't know . . ."

"Is that even allowed?"

I'd hardly spoken to anyone on the bus that day, but I took off my earphones and yelled down the aisle, "The man flew the fucking space shuttle. I'm pretty sure he can handle a rent-a-jet on a thirty-minute flight to Miami."

I lost. We would spend the night at a Renaissance near the airport attached to a mall called the International Plaza. At the bar of the Capital Grille, I ordered a dry-aged porterhouse and three glasses of a Napa Valley pinot before the bartender announced last call. I walked back to the hotel nice and buzzed, weaving through a series of outdoor walkways made to look like a street in *Roman Holiday*, but with sports bars and a California Pizza Kitchen. Stuck in the great coastal shit hole of Tampa Bay, and for the first time in months, I was free.

39

The Bed Wetters

WASHINGTON, DC → ATLANTIC CITY → NEW YORK CITY, SUMMER 2016
Democrats started to worry. When Trump called Bill Clinton a rapist on Fox News, Hillary responded by telling CNN, "I have concluded he is not qualified to be president."

Brooklyn was elated that she'd really *gone there*.

"She actually called the presumptive Republican nominee unfit," Hired Gun Guy said.

"Uh, yeah, but he did call her husband a rapist," I said.

When Trump suggested he'd reduce the national debt by negotiating with creditors to accept less than a full payment, one of his more batshit ideas that would lead to global economic calamity, Brooklyn issued a press release calling the proposal "risky."

Hillary was still following the Mitt Romney Playbook, not realizing that she was the Romney in the race. She tried to make Trump a cold corporate titan who got rich screwing over the little guy. She had an event in Atlantic City with the fading outline of the Trump Plaza casino sign in the background.

"Just down the boardwalk is the Trump Taj Mahal. Donald once called it the eighth wonder of the world," Hillary said. She liked trolling Trump and smiled a little as she gestured toward the decrepit hotel, now dotted with busted light bulbs and dust-covered faux-gold urns. She tried to make her put-downs snack-size. "People get hurt and Donald gets paid."

But Trump wasn't Romney. He could have a car elevator and no one would care. Reporters would ask for rides on the damn thing. Meanwhile, Trump feasted on the FBI press conference and his rat-a-tat recitation of "Crooked Hillary," while she criticized his plan to eliminate the estate tax and cut corporate tax rates.

"Sometimes you get the feeling they're in a professional boxing match and he's in a street fight, and they're coming in with their gloves on," Rev. Al Sharpton said. "This is a street fight with a guy with a razor and a broken Coca-Cola bottle. You've got to fight him like that."

Chuck Schumer told Jonathan Martin that he'd implored Brooklyn to hire a senior staffer whose only job was to respond to Trump on an hourly basis. John Hickenlooper, the Colorado governor, said he was reading *The Art of the Deal* and that Hillary "has to be careful because now he [Trump] has momentum."

Matthew Dowd, a former chief strategist to George W. Bush who is now an independent, told me in late February, "Hillary has built a large tanker ship and she's about to confront Somali pirates."

Brooklyn blew it all off. The math was on their side. "It wouldn't be a general election without some early bed-wetting from Washington insiders," Robby said.

NO CALLER ID flashed on my phone. I'd left the newsroom and was sitting in Bryant Park to soak up the early summer air and clear my head. It was June, days before Hillary would win the nomination. People with normal jobs spread picnic blankets and wine and Brie out on the lawn as a trio of flamenco guitarists set up on a temporary stage.

"Hello?"

"Amy, it's Donald Trump . . ."

In a speech in San Diego, Hillary had finally delivered the takedown Democrats wanted to hear, mixing in her wit and signature sarcasm. Megan Rooney, who like Dan Schwerin worked with Hillary at the State Department, perched on the armrest of Hillary's luxe leather seat on the flight to San Diego to hammer out the most lacerating lines.

Hillary looked like she was enjoying herself when she called Trump's foreign policy pronouncements "not even really ideas, just a series of bizarre rants, personal feuds, and outright lies."

"This is not someone who should ever have the nuclear codes because it's not hard to imagine Donald Trump leading us into war just because somebody got under his very thin skin," she said.

Trump had called the *Times* during the speech to yell into the phone, "I'M NOT THIN-SKINNED AT ALL. I'M THE OPPOSITE OF THIN-SKINNED."

The day before, Policy Guy gave me a rundown of what the speech would say so that I could write a curtain-raiser. When Carolyn came back from the page-one meeting with the most beautiful words in the English language, "They want it for the front," I reached out to the Trump campaign for comment. I expected a statement from Hope Hicks, Trump's competent and responsive spokeswoman. Instead, Trump called directly.

In this period, most of my colleagues had stories of standing in line at Starbucks or climbing onto the elliptical when the infamous "NO CALLER ID" Trump call came in. I'd spent months requesting interviews with Hillary. Always the answer from Brooklyn, no matter how positive or substantive the topic, was either stone-cold silence or a hard no. But there I was in Bryant Park, picking up my phone to . . .

"Amy, it's Donald Trump . . ."

I dug around in my bag for a pen and pulled out some loose scraps of paper. Trump repeated the phrase "America First" at least six times, attributing his favorite pet phrase to "your very good, very smart colleague David Sanger, excellent guy." (I agreed.) He then laid out his plan to counterattack.

"Bernie Sanders said it, and I'm going to use it all over the place because it's true," Trump said. "She is a woman who is ill-suited to be president because she has bad judgment."

We bantered about *The Apprentice* a little. ("Can you believe Schwarzenegger thinks he can do it?") Then I said something I never should have said.

"Thanks very much for calling, Mr. Trump. I've been covering Hillary since 2007, and she's never called me."

"Is that right?" The wheels were turning. "When was the last time she talked to you?" Trump asked.

I thought about it. "I don't know. I guess it's probably been five, six months since she had a press conference."

Silence. The wheels turned some more.

"You know why?" Trump said. I wanted to say, Yes, Mr. Trump, because she hates us and thinks we have big egos and tiny brains. But I'd already said too much. "She doesn't have the stamina," Trump said. He raised his voice. "It takes STAMINA to talk to the press."

I don't know if I gave Trump the idea or he'd had it for weeks, but after that he started to tell crowds, "So, it's been two hundred and thirty-five days since Crooked Hillary has had a press conference . . ." His campaign started to blast out a daily reminder: HILLARY HIDING WATCH: DAY 262 SINCE LAST PRESS CONFERENCE.

40

Off the Record . . .
Until Hacked

PHILADELPHIA, JULY 2016

An oppressive heat wave settled on Philadelphia the week of the Democratic National Convention. The city wasn't baking like south Texas, where the temperatures climb to over 100 and sit stubbornly on top of strip-mall parking lots, but the air was thick and soupy. Even the breeze that came in before a downpour carried an ominous combustibility.

Russians had hacked into the Democratic National Committee emails. The Bros continued with their death threats. They'd jumped on an email exchange in which a DNC press aide said I was "kind of friendly" with Debbie Wasserman Schultz, who resigned over the hack. The Bros took this as clear evidence I was colluding with the chairwoman to elect HRC, texting me their usual short and sweet messages like "You Hillary loving cunt." My emails to the DNC were nothing. Mostly Debbie and I talked about dieting, workouts, and dealing with our unruly Jewfros. Not exactly the makings for a grand conspiracy. But *Times* reporters, Democratic leaders, donors, Brooklyn aides, hell, even campaign volunteers, were panicked. Could our emails be next? Sources got so nervous at least two of them started to text me "OTR until hacked."

EVER SINCE THE State Department, Hillary had a thing about Putin and vice versa. Months before she declared her candidacy, at a paid speech in Winnipeg, she'd broken into an impersonation of Putin.

Without being asked about the Russian leader, she put on a deep, accented voice and swayed her head left and right pretending to be Putin having a conversation with himself. "Vladimir, you think you'd like to be president again . . ."

Now, as she tried to woo Republicans, Hillary sounded positively Reaganesque as she criticized Trump's Russophilia. "He praises dictators like Vladimir Putin and picks fights with our friends," she said the previous month. "Putin will eat your lunch." In May, she said a Trump presidency would be like "Christmas in the Kremlin."

But the evidence that Russian intelligence had been behind the twenty thousand stolen DNC emails released at precisely the right time to disrupt the convention handed Brooklyn a good vs. evil plot out of a le Carré novel.

"There have been larger forces at work to hurt Hillary Clinton," Robby said in the opening hours of the convention.

I wrapped all this Russian intrigue into a scene-y blog post casting Trump as a Manchurian candidate, Putin's unwitting puppet. I liked the story, but I also saw the wisdom when my editors deemed it was too unsupported to publish.

"GOD FORBID THEY make one dick soft in a swing state!"

Linda Bloodworth-Thomason talked as you'd expect the creator of *Designing Women* to talk. She was particularly saucy the second morning of the convention. She'd been fighting with Brooklyn over a biographical video she and her husband, Harry Thomason, made that was supposed to introduce Hillary that evening.

At the 1992 convention, the Thomasons, Hollywood filmmakers and writers who have been friends with the Clintons since the 1970s, wove Bill's hard-knocks childhood into one of the greatest origin stories of modern politics with *The Man from Hope* video. "Some people think that Bill must have been born wealthy and raised wealthy," Hillary says while sitting on a patio in pink shoulder pads. *Insert Dickensian images of his childhood in rural Arkansas here.* "Well, you know instead of being born with a silver spoon in his mouth, he was really born in a house with an outhouse in the backyard."

The Thomasons wanted to pull off the same feat for Hillary, but her origin story—a nuclear family in an upper-middle-class Chicago suburb and Wellesley education—didn't exactly give them the same artistic tableau. They needed to go one generation back to Hillary's mother, Dorothy, abandoned by her parents at eight years old, on her own and working as a housekeeper for three dollars a week at fourteen, before raising the little girl who was one election away from becoming the FWP. The Thomasons enlisted Meryl Streep to narrate. "Her name, like another little girl who got caught up in her own history-making tornado, was Dorothy . . ." Streep said. The footage then moved on to the women who came before Hillary—Harriet Tubman, Susan B. Anthony, Eleanor Roosevelt, Rosa Parks, and Sally Ride. It ended on a 2010 photo of Dorothy, Hillary, and Chelsea at Chelsea's wedding.

On my second day in Philly, I sat in a cracked leather armchair in the lobby of the Hilton Garden Inn to download the ten-minute video. It gave me the chills.

The video, called *Shoulders*, was the first time so far at a convention designed to celebrate Hillary when I actually felt moved by her arc, not the one created for her by pollsters and strategists, but the real one, rooted in a daughter's love for her mother. I know that sounds cheesy, but Meryl was narrating . . .

Ad man Jim Margolis, Joel, and Mandy took one look and killed it.

The plan had been to *not* overplay the history-making part. Hillary did that already in the Brooklyn Navy Yard speech and in the where-are-all-the-men introductory video. Now she needed to broaden her appeal and reach Republicans and men. Brooklyn decided Hillary would promise to "break down all barriers." For Christ's sake, no glass ceilings. "Press will immediately go to this as a more narrow attempt to appeal to women," Joel said when the "ceiling language" had crept back into a speech during the primaries. Podesta agreed, "Ceilings language is catnip for press and will drown out break down barriers."

Like a lot of Hillary's female friends, Linda worried the campaign could've been slicing and dicing the data needed to sell any Democratic nominee. The newer aides didn't know Hillary and didn't really care to. She was a blank slate on which to project a rainbow coalition of Obama voters. "Dorothy is the only poetry she's got," Linda said.

The campaign went instead with a biographical video from Shonda Rhimes, the *Scandal* creator. (All roads lead back to Tony Goldwyn.) On the final night of the festivities, the video would show scenes of Hillary when she was a New York senator, comforting first responders after the September 11 terrorist attacks and imploring George W. Bush in the Oval Office for $20 billion to rebuild New York. In this version of her story, Hillary sits in a sunny kitchen, in front of a vase of fresh flowers and a stack of cookbooks (wink, wink, women) as Morgan Freeman's languid narration explains how she almost single-handedly captured and killed Osama bin Laden.

"Look at her, look at her face," Freeman says of the photo of Hillary in the Situation Room holding a hand to her mouth and surrounded by men paralyzed, their eyes fixed to a screen showing images of a Navy SEAL raid on a compound in Pakistan. "She's carrying the hope and the rage of an entire nation."

In 2011, Hillary attributed her expression in the photo to her pollen allergies. But I guess "the hope and rage of an entire nation" sounded better.

I WALKED PAST the Ritz-Carlton near Centre Square, where black SUVs lined up outside to drop off wealthy donors. They wore a rainbow of credentials and VIP passes around their necks. Protesters yelled into bullhorns. And each time the bellmen opened the door and welcomed guests into the Ritz's cool marble lobby under the calming hum of classical music, protesters' chants of "Hell no, DNC!" and "We won't vote for Hill-ah-ree" and "Wik-EE-Leaks! Wick-EE-Leaks!" would blow in with the soup and retreat again as the doors swung shut.

"This is what democracy looks like!" a protester with dreadlocks and Clearasil-perfect skin yelled from a bullhorn. A security guard pushed along two young girls who wore T-shirts that said I'M NOT A SUPER PRED-ATOR. The sun beat down on the cement, and the police stood in clusters trying to find shade.

41

The Red Scare

"I saw Debra Messing downstairs earlier," a woman with Prada sneakers said to no one in particular as the elevator doors closed. The blue-lit lobby at the Le Méridien had leather benches and a sleek silver bar where a couple of donors sipped mimosas before stepping out into the soup.

"Can you hit six for me?" the woman said.

When the elevator doors opened again, Bryan Cranston got off.

This is what democracy looks like.

The press took our seats in a ballroom at the Le Méridien for a "Bloomberg Politics Breakfast with John Podesta." I walked the length of the rectangular table twice, passing name tags of journalists who didn't need any introduction—Andrea Mitchell, John Heilemann, Susan Page—before I remembered Jonathan had RSVP'd for the breakfast and then given the invite to me. I pulled out the chair behind JONATHAN MARTIN, THE NEW YORK TIMES.

Podesta sat in the center, hoisting his head left and right as the reporters yelled, "John, over here!" and "John, to your left."

"John, tell us a little bit first about last night. About I think it's safe to say eight, eight thirty, there was a lot of tension on that floor . . ."

"John, are there still more disruptions, more booing, more anger today?"

"John, what if there is a walkout, what if there is a big protest? How would you deal with that?"

"Everybody is going to get a chance to vote, and then she is going to make history," Podesta replied.

Next question.

When the talk started to lag, I saw an opening. The Russians hadn't come up. The Bernie Bros seemed the more imminent threat.

The Russian meddling subplot still seemed far-fetched. Even Nancy Pelosi, when asked if the Trump campaign could've played any role in the release of the twenty thousand DNC emails, said, "I have no reason to think that." Hillary aides pushed that the Russians were helping Trump, but always in hushed off-the-record winks and nods, like they didn't really believe it but wanted it to be true because it would help Hillary. "He certainly has a kind of bromance going on with Mr. Putin, so I don't know . . ." Podesta said.

When the *Washington Post*'s Phil Rucker asked Robby if he was concerned about threats Julian Assange had made about future hacked emails that WikiLeaks could release to inflict damage on Hillary and the campaign, Robby replied, "I'm not going to put a lot of stock into what Julian Assange says. I mean—he says a lot of things and, um, so I'm not going—I'm not going to pay attention to that."

I asked Podesta, who'd worked extensively on cybersecurity in the Obama White House, whether his own or anyone else in the campaign's emails may have been hacked.

"There is—we don't know the answer to that, and we feel like we have robust security within the campaign, and obviously are contacting on a daily basis," he said.

"So the FBI hasn't warned you or anyone else—" I continued.

He cut me off. "No. The FBI, no."

His head whirled left, right. Next question. "Andrea?"

A few days later I saw the Yahoo News headline FBI WARNED CLINTON CAMPAIGN LAST SPRING OF CYBERATTACK. FBI agents had met with senior campaign officials in Brooklyn to warn them that their emails, and particularly Podesta's, may have been penetrated through so-called spear-phishing emails. I never knew why Podesta hadn't been honest about that. But maybe it was because Outsider Guy had been right when he told me, "No one takes you seriously." If I'd been Jonathan Martin, would Podesta have at least pulled me aside afterward to explain things?

Brooklyn didn't want any more hacking news to overshadow the convention, but Trump couldn't resist urging Russian intelligence to hack into Hillary's emails. "Russia, if you're listening, I hope you're able to find the thirty thousand emails that are missing," he said at a news conference in Doral, Florida. "I think you will probably be rewarded mightily by our press."

Thing was, Trump didn't need to put on a public display. In March, around the time the FBI paid a visit to Brooklyn, Russian hackers sent a phishing email to Podesta's personal Gmail account. He'd opened the link and clicked on the Change Password button.

42

Gladiator Arena

PHILADELPHIA, JULY 2016

While Trump had emerged as a fog-enshrouded silhouette at the Republican National Convention in Cleveland, Hillary, a reluctant star in the four-day infomercial of her life's work, had watched most of her convention at home on TV. She didn't arrive in Philly until Wednesday and didn't appear onstage until the third night and only for a second to give Obama a warm, prime-time hug.

"She's an introvert," Tom Vilsack, the agriculture secretary, told me. "The spotlight is pretty glaring, and she likes to deflect from it."

Brooklyn worried, too, that the hagiography of a political convention would only add to the perception that Hillary is all about herself. They made behind-the-scenes efforts to revise "I'm With Her," to the less self-centered "She's With Us."

Hillary had no problem with conventions when she played a supporting role. In 1992, she'd waved and exhaled clutching Chelsea's hand tight, a tinge of relief and anticipation and a what-now look on her face as the confetti blanketed Madison Square Garden moments after Bill, still doughy and cherub-cheeked, spoke of the "New Covenant" and a country "of boundless hopes and endless dreams"—a far cry from the 2016 message: *Vote Hillary. The other guy is a sociopath.*

SHE'D HARDLY FINISHED delivering her nomination acceptance address when everyone agreed that the speech had been "serviceable." Editors and reporters nodded that she "did what she needed to do." She hadn't blown anyone away, but she just needed "to not be Trump."

A month earlier, Hillary, who'd begun jotting down what she wanted to say after the New York primary in April, sat down with Policy Guy and speechwriters Dan and Megan. Dan had heavy brows and a neat beard that ran directly into his head of brown curls. In addition to being Hillary's chief speechwriter, he was best bros with Brown Loafers, which protected his position even as outsiders constantly grumbled that Hillary's speeches had become a blur of bromides. If Hillary wasn't promising to "build ladders of opportunity," she was vowing to "reshuffle the deck" and to make sure "every child can live up to his or her God-given potential."

I didn't envy Dan. I had three, maybe four, editors read my big stories. He had to collect and entertain opinions from an apparatus as ungainly and hierarchical as the Clinton campaign (plus family) and somehow weave it all into a salient speech. Dan's emails often included feedback like "Big revision from WJC." Or Podesta's critique that an early draft "has the feel of the kitchen sink being thrown in." Joel provided the constructive: "For what it's worth, I hate the yelling/screaming/Trump graph. It sounds completely contrived and doesn't deliver a big joke or a big point." Then there were the findings of focus groups, like "Let's make sure we don't go too far with anti-corporate rhetoric despite the anger we heard in yesterday's groups in Detroit." I imagined an elderly woman behind two-way glass at an office park in greater Miami reaching for the chicken chow mein and explaining to JFK speechwriter Ted Sorensen, "That 'Ask not what your country can do for you . . .' line just seems so bossy."

Dan's position as the *chief* speechwriter irritated some of Hillary's aides who projected onto the Andover grad all their frustrations that crafty men always seemed to upstage more talented, low-key women, namely Megan. Dan had already started to think about Hillary's inauguration speech and would talk about how he'd structure his White House speechwriting team, with everything (national security, domestic agenda, etc.) reporting to him.

Lissa Muscatine, the warm co-owner of the Politics and Prose Bookstore in Washington who had written speeches for Hillary since the White House years and knew her and her voice better than any of the younger generation of writers, pitched in, as did Jon Favreau, Obama's star speechwriter. The Guys hated it when we gave Favreau credit for a Hillary speech. Favs brought out all their insecurities about 2008 and the Obama bros. Yes, in 2008 Jon had posed groping the right breast of a cardboard cutout of Hillary. But mostly Jon's involvement undermined Dan, who considered himself the New Favreau.

Dan and Megan and Policy Guy ordered room service in a deluxe king suite at the Logan hotel and worked until four of the final morning of the convention.

In a backstage locker room, Hillary scratched last-minute changes to the final draft right until she walked onstage. In a white Ralph Lauren pantsuit and crew-neck shirt underneath, she looked resplendent. Seeing Hillary take the stage as the Democratic Party's nominee, the waving white solo speck at the center of a circumference of fifty stars, made all the complicated feelings I had about her briefly fall away.

She warned of "a moment of reckoning"; referenced great presidents past, both Republican and Democrat. "He's taken the Republican Party a long way from 'Morning in America.'" And "The only thing we have to fear is fear itself." She scattered in some self-deprecation—"It's true, I sweat the details of policy." And she recited some résumé points: "I went to a hundred and twelve countries." "In the White House Situation Room . . ." And "I wrote a book called *It Takes a Village*."

THE PRESS STARTED to file out of the arena leaving the *Times'* row of assigned seats littered with empty Pepsi cans, tangy crumbs from a family-size bag of Cheetos, and the wrapper from a frosted Pop-Tart. The intensity of covering the convention had caused us all to revert back to eating like ten-year-olds reared in 1980s suburbia. I stayed behind and kept my eyes fixed on the stage. I remember feeling delight as I watched Hillary. She'd looked childlike, staring up with an oblong mouth and the bulbous, happy eyes of an anime character as the red, white, and blue balloons and a deluge of patriotic confetti showered around her. I saw her kick a red balloon with her kitten heel and catch an oversize blue one with white stars in her arms, hugging it tight as if her night had been oxygen trapped inside its latex belly.

It was after midnight when Pat Healy and I closed our story. Carolyn, putting an end to two weeks straight of convention coverage, picked up one of the half dozen cans of Diet Coke and, noticing that it was empty, dropped it and picked up another. She stared at our story on her screen in its sixty-four-point font, as Very Senior Editor looked over her shoulder. When she finally hit Send, dispersing our copy to printing plants nationwide, I packed up my temporary desk, including all my chargers and notebooks.

THE NEWSPAPER THAT next morning was a thing of beauty. The headline CLINTON WARNS OF MOMENT OF RECKONING—ACCEPTS HISTORIC NOMINATION, PROMISING TO "REPAIR THE BONDS OF TRUST" stretched across all six columns. I stared at my name in print under Pat's and next to a photo of Hillary waving to the convention crowd.

> PHILADELPHIA—Hillary Diane Rodham Clinton, who sacrificed personal ambition for her husband's political career and then rose to be a globally influential figure, became the first woman to accept a major party's presidential nomination on Thursday night . . .

I'd thrown my elbows around enough that by then I wrote or cowrote front-page stories several times a week, but this one was different. I held the paper in my hands staring at my oddly Slavic surname. The name passed to me from the shtetls of Poland and the Lower East Side to my traveling salesman grandpa in Waco. The name I just couldn't change when I married an Irishman. I envisioned mothers reading our lead to their daughters, historians referring back to our words.

I'd been on the beat for so long that it hadn't occurred to me that there would be a time when Hillary would no longer be running for president. With the conventions behind us, everyone started to think about what they'd do next. This added another layer to our already competitive politics pod. I had colleagues like Jonathan Martin who were born to cover politics. A couple of years back, when I told Jonathan I was in Texas for my grandpa's funeral and that he was almost 101, Jonathan wrote back immediately, "Wow . . . Born in the Taft admin!" I told him he could own any beat. "Nah," Jonathan replied. "I'm like a podiatrist. I just do toes." But I'd become a campaign reporter not because of my love for politics or my encyclopedic knowledge, but because my career had from an early age become intertwined with Hillary's. Now I couldn't imagine a future separate from Hers. I didn't really want to move to Washington, but I needed to be there for the Final Act— covering Hillary as the FWP.

"So, I know I was wavering on DC the last time we talked, but I really would love to cover the White House if Hillary wins . . ." I told Very Senior Editor over coffee that morning.

He crumpled his eyebrows behind his wire-rimmed glasses. "Would you move to Washington?" he asked.

I nodded, hoping to convey that I would do *anything* for the *Times*. "Absolutely," I said.

Bobby never said so, but I knew he would've preferred that my next posting take us to Hong Kong or Delhi, somewhere far away from the Clintons and that reminded us of the adventures we'd had in Japan. But he knew me well enough to know that I wouldn't be able to let Hillary go if she became the FWP. In the past few weeks, he started to peruse job listings in DC and signed up for Trulia updates, fixer-uppers in Georgetown mostly.

"Well, that would make sense," Very Senior Editor said. He was supportive but noncommittal.

Maybe I wanted my editors to want me for the White House job more than I actually wanted the job.

Several years back I'd let one of the Iraq War vets who taught my boot camp talk me into signing up for something called a Spartan Super Race. This included an eight-mile run over sand and mud, crawling under barbed wire, and jumping over fire. When I arrived in Staten Island to register, I looked around and saw brawny shirtless men in military fatigues writing their race numbers on their foreheads with permanent marker. (I knew I wasn't cut out to be a Spartan when I inquired whether this numerical system would make my skin break out.) Five hours later, when I got to the grand finale—a gladiator arena where race organizers in leather loincloths pummel contestants with pugil sticks—I didn't feel like a badass Spartan queen. I felt really fucking stupid for signing up in the first place.

The White House job would've been Gladiator Arena. Four more years of fighting with The Guys. Four more years of fighting to write the same stories as the rest of the pack. Four more years of waiting for a press aide to tell me "OFF THE RECORD" and "FOR PLANNING PURPOSES ONLY" that I needed to arrive by 7:00 a.m. so that I could sit in a press van. But I had some masochistic urge to do it anyway.

After all, I'd jumped through fire—or at least being fired—to get there. Condé Nast had done away with the rover program years ago, but part of me was still the interloper staring at the back of David Remnick's head in the elevator. I still needed to prove something. Christina, one of my few friends who can get away with passing off unsolicited life advice picked up from her therapist and self-help books, diagnosed me as suffering from "striveritis."

AFTER THE LAST balloon had dropped, Hillary spent at least an hour backstage, exchanging air kisses and pleasantries with Meryl Streep, Katy Perry, and other assorted VIPs, before she went back to the Logan hotel and sipped champagne with friends. Trump had entered the third day of his feud with the grief-stricken Muslim parents of Humayun Khan, an army officer killed in the Iraq War, and a lot of Hillary's friends saw an opening.

"Now is your chance, Hillary, to get out and show people who you really are. They'll love you like we love you," one girlfriend told Hillary backstage.

But Hillary was done trying. A week earlier, she'd cut off Joel and the pollster John Anzalone, as they walked her through the almost daily reminder that half the country disliked her. "You know, I am getting pretty tired of hearing about how nobody likes me," she said.

"Oh, what's the point? They're never going to like me," Hillary told this friend.

I thought of something Bill had told me in Uganda, how Hillary had laughed when they were dating at Yale and he suggested she run for office. "Look at how hard-hitting I am," she'd told him. "Nobody will ever vote for me for anything."

After the Philly rally, Hillary, Bill, Kaine, and his wife, Anne Holton, would embark on a three-day Jobs Tour across an industrial swath of Pennsylvania and Ohio. Brooklyn thought Trump's meltdown with the Khans, and all the independents and military men who spoke at the convention, could give Hillary an opening with Romney voters.

"I'm also going to pay special attention to those parts of our country that have been left out and left behind," Hillary had said in Philly. "From our inner cities to our small towns, from Indian country to coal country. From communities ravaged by addiction and places hollowed out by plant closures. Anybody willing to work in America should be able to find a job and get ahead and stay ahead. That's my goal."

But after the bus tour, Hillary spent most of August hobnobbing with the ultrawealthy. She hauled in $143 million that month, including $50 million at twenty-two fund-raisers in the last two weeks of August. After months of trying to portray Trump as the embodiment of "a system where the rich and powerful stick it to everybody else," Hillary closed out the summer by averaging $150,000 an hour.

43

"HRC Has No Public Events Scheduled"

Hillary always broke down Trump supporters into three baskets:

- BASKET #1: The Republicans who hated her and would vote Republican no matter who the nominee.
- BASKET #2: Voters whose jobs and livelihoods had disappeared, or as Hillary said, "who feel that the government has let them down, the economy has let them down, nobody cares about them, nobody worries about what happens to their lives and their futures."
- BASKET #3: The Deplorables. This basket includes "the racist, sexist, homophobic, xenophobic, Islamophobic—you name it."

The Deplorables always got a laugh, over living-room chats in the Hamptons, at dinner parties under the stars on Martha's Vineyard, over passed hors d'oeuvres in Beverly Hills, and during sunset cocktails in Silicon Valley. As Hillary was breaking down the baskets to donors, Trump spent August as if anticipating the gaffe gift basket. He told a crowd in Manchester that Hillary "lies and she smears and she paints decent Americans—you—as racists."

I'll pause here to add a subcategory to Brown Loafers' Gaffe Matrix: the damning statements that aides see coming but don't have the cojones to stop.

In 2008, Michelle Obama's team grimaced every time she said, "For the first time in my adult life, I am really proud of my country," at dozens of private fund-raisers before she blurted it out at a Milwaukee rally and handed the GOP an attack on the Obamas' patriotism more potent than any flag pin. Mitt Romney didn't start talking about the "47 percent" that night in Boca Raton. The Guys and other senior aides who heard Hillary talk about the baskets could've warned her that the whole Deplorable thing wouldn't look so good if it got out. But Hillary liked the line. Who was going to tell her to stop?

The Travelers weren't allowed to cover Hillary's fund-raisers. This made her comfortable testing out new lines in the safe haven of friends willing to pay $10,000-plus. We traced the origins of a general-election favorite—"Friends don't let friends vote for Trump"—to Hillary's LGBT fund-raiser with Cher in Provincetown.

"I stand between you and the apocalypse!" a line that Hillary used in interviews throughout the fall and that ended up a headline in a *New York Times Magazine* cover story started in the backyard of Jimmy Buffett's beachside estate. Calvin Klein, Harvey Weinstein, Andy Cohen, and the other guests, who'd paid $100,000 to attend, ate up the apocalypse joke.

For the "Hey Jude" finale, Hillary danced under a tent lined with tiki torches alongside Buffett, Jon Bon Jovi, and Paul McCartney. To see Hillary dance, really dance and not attempt the whip/nae nae on *The Ellen DeGeneres Show*, was worth the steep price tag. She is a collection of mismatched limbs and hip rolls, the joyous, intoxicating cavort of a rhythmless but happy woman.

For months on almost daily conference calls, Joel told Hillary that most of the country hated her. "Note to self: Tho she professes indifference to what people think, she is acutely aware of opinion polling on her," Diane Blair wrote of Hillary during the White House years. In the Hamptons, Hillary felt loved. Hillary the introvert, Hillary the "compassionate misanthrope," Hillary of Coke-bottle glasses and hairdo changes became the belle of the ball.

I stopped traveling regularly. Hillary's public schedule was scant and while I've often surrendered my dignity for the job (yelling at Bill

Clinton about emails as he visited with impoverished children comes to mind), I drew the line at sitting outside rich people's houses hoping for a glimpse of Hillary. On most days, the Travelers didn't see her. Typical dispatches from the road would read, "Pool was unable to hear a word of HRC's remarks at first cocktail party. She also took questions, which pool was also prohibited from hearing," and "Pool could hear Clinton's voice, but could not make out any exact words."

The caretaker of a Bridgehampton compound spotted the press sitting around in the pool house and asked when they'd go inside to see her. "We told him no, we aren't allowed to do that. He seemed incredulous," Peter Nicholas wrote. "Pool reported no sighting of Clinton the entire day." Annie Karni summed up the summer as a "sensory deprivation experience."

After our last interview, Hillary promised to compare our reading lists at the end of the summer. I nagged and nagged the campaign, even preemptively sending my own list which included not one but two self-help titles. But this never happened. Instead, Hillary chose to talk about her reading list with an eleven-year-old blogger at *Elle* magazine.

Each week I forwarded to editors the "FOR PLANNING PURPOSES ONLY" guidance from Sarah, who fielded the Travelers' complaints from the Brooklyn HQ:

Saturday, 8/20—HRC will attend finance events on Nantucket, MA and on Martha's Vineyard, MA.
Sunday, 8/21—HRC will attend finance events on Cape Cod, MA.
Monday, 8/22—HRC will appear on *Jimmy Kimmel Live!* and attend finance events in Los Angeles, CA.
Tuesday, 8/23—HRC will attend finance events in the Los Angeles and Orange County areas and then the Bay Area, CA.
Wednesday, 8/24—HRC will attend finance events in the Bay Area, CA.
Thursday, 8/25—HRC will campaign in the Reno, NV area.
Friday, 8/26—HRC has no public events scheduled.
Saturday, 8/27—HRC has no public events scheduled.

Sunday, 8/28; Monday, 8/29; Tuesday, 8/30—HRC will attend
finance events in the New York [area].

Wednesday, 8/31—HRC will address the American Legion's 98th
National Convention in Cincinnati, OH.

Thursday, 9/1—HRC has no public events scheduled.

Friday, 9/2—HRC has no public events scheduled.

Saturday, 9/3—HRC has no public events scheduled.

Sunday 9/4—HRC has no public events scheduled.

"Does she even want to be president?" an editor wrote back.

I had so much time in New York in late August that I started thinking about my ovaries again. In the three years since I'd blown off Dr. Rosenbaum's advice to have a baby and bring an au pair on the campaign trail, Bobby and I had filed the baby thing so deep in the back of our minds you'd think I was a seventeen-year-old on the pill. I had no eggs on ice. We didn't talk about becoming parents or envy our friends' kids. Our five-year plan included potentially moving to Washington, followed by the 2020 election.

As the taxi sped along the East River on my way to an appointment with a new gynecologist, I looked out the window and noticed Roosevelt Island, where Hillary had delivered her milquetoast kickoff speech. The island, the bust of FDR where it all started, taunted me. *Your fertile window is closing.*

Dr. Broderick looked at my file. She had a moon face framed by wisps of strawberry-blonde hair. "You're not old," she said. I liked her.

I told her about my insane work schedule ("So, I can't do anything at least until November, but then there will be the transition . . ."). I told her about the judgy fertility app that kept guilting me while I was on the road. She agreed with Maggie that I should delete it immediately. I said that if Bobby and I decided to really do this, it would have to be an "off-cycle baby," what the female Travelers said when they would time births between presidential campaigns.

Dr. Broderick listened. She didn't look at me as though I was crazy, which made me seriously question what New York women in their midthirties have confided to her. She said we'd run some tests when I had time, and then said, "It's New York, Amy, everybody who wants a baby gets a baby."

IN THE OLD days, before David Geffen slept in the Lincoln Bedroom (twice), before a Texas oilman and a Cincinnati trial lawyer got rides on Air Force One, and before a telecom executive and her third husband honeymooned in the White House, Democrats raised money from labor unions and East Coast liberals. Then Bill and Hillary came along and the game changed.

With the help of moneyman Terry McAuliffe, the Clintons built a coast-to-coast network of Southern trial lawyers, Hollywood producers, Midwestern businessmen, and Wall Street executives, and they treated these donors like family.

Hillary showered them with handwritten notes and birthday calls ("It's your secretary of state calling to wish you a happy birthday . . ."). Terry once left his wife crying in the car with their newborn son so he could pop in at a fund-raiser on the way home from the hospital. "I felt bad for Dorothy, but it was a million bucks for the Democratic Party," he wrote. Donors ate up the personal treatment. One donor who loved the attention complained to me that Obama didn't do the same. "I could donate a kidney to Malia and not get invited to the White House," he said.

Hillary was always paranoid about money. She once said she'd been worrying about her commodity trades while in labor with Chelsea. And just as leaving the White House with $5 million in legal fees drove a "dead broke" Hillary to hit the Wall Street speech circuit, lending her 2008 campaign $13 million of her own money had turned Hillary in 2016 into both a cheapskate and a ravenous fund-raiser. But the Clintons' style of fundraising was anachronistic, especially in a "Two Americas" election year.

Throughout the primaries, Hillary would exhaust herself dropping off the campaign trail weekly to crisscross the country and hit up a living room full of rich people for $2,700 checks. Bernie, meanwhile, had a couple of kids working his website who turned online donations of five dollars or ten dollars into a $230 million windfall. His campaign raised $8 million in the forty-eight hours after winning New Hampshire, $3.6 million in the days after #BirdieSanders.

Hillary relied on Dennis Cheng, a sleek London School of Economics graduate who wears pocket squares and Paul Smith socks and has amassed the most impressive Rolodex in Democratic politics. He'd leveraged the precampaign period to build up a $250 million endowment for the Clinton Foundation—or what one former aide called "Chelsea's nest egg." Dennis struck a more refined, reserved presence than the avuncular frat boy Terry, but they shared the same attentiveness.

Donors described Dennis as "a master concierge." He offered a full menu: For $250,000 per person, donors could have an intimate chat with Hillary at a waterfront estate in Sagaponack. For $2,700 the children (sixteen and under) of donors at the Sag Harbor estate of hedge-fund magnate Adam Sender could ask Hillary a question. "I go to Dalton, but how would you make sure every kid gets to go to a good school?" asked an eight-year-old boy, who in ten years will surely be dropping the H-bomb into mid-conversation. Want a family photo with Hillary? That's an extra $10,000.

I'D CARVED OUT a tabloidy mini-beat reporting on Bill and Hillary in the Hamptons, partly because any story that merged the two topics almost guaranteed me the cover of Sunday Styles. But I also had some sociological urge to understand why the area and its old money had such a hold on us transplants.

Trump and the Hamptons never mixed. His is not the kind of wealth that wants to hide behind the hedges. He'd once yelled at Rupert Murdoch that he'd sue for libel after the *New York Post* reported that the exclusive Maidstone Club in East Hampton denied his membership. Screw 'em, Trump would take the gaucheness of Palm Beach instead.

But Bill and Hillary were Gatsby's pursuers.

I didn't believe the spin that Hillary only spent time in the Hamptons because she *had* to fund-raise. She thought she was going to win and she liked it there. Those were her people.*

I discovered the Hamptons after Mr. Ascot at Condé Nast passed off that unwanted invitation to the *Sex and the City* party. But for Bobby and me, the exclusive shores of Long Island usually meant a half day of fluke fishing in Montauk and a room at Daunt's Albatross, a 1970s-era motel with a complimentary VHS library and walls so thin that we played guess-the-fake-orgasm. If the motel didn't work out, we did what Bill, Hillary, and most of New York did: We mooched off rich friends.

In late August, Bill showed up with Hillary's two dogs, Maisie, a curly-haired mutt, and Tally, a toy-poodle mix, to crash in Steven Spielberg's East Hampton guesthouse. Previous summers, when the Clintons rented their own beachside estates, Hillary's brothers, Tony and Hugh, and the entire extended family showed up—the moochers of the moochers. (A grocer in East Hampton told me he saw Roger Clinton buying milk in a tracksuit.) In 2012, the Clintons had a dispute over the security deposit of the twelve-thousand-square-foot East Hampton mansion they rented, the one with the heated pool that typically goes for $200,000 per month in August.

These details went over about as well as the Yorkie and only added to The Guys not taking me seriously. But I couldn't help it. I mean, who brings their dogs?

"WHERE HAS HILLARY CLINTON BEEN?" ASK THE ULTRARICH was just about the last front-page headline that Hillary wanted splashed across

---

* At least, Hillary thought they were her people until she took their money and lost to Trump. I'll never forget sitting in the Upper East Side home of one of Hillary's most loyal Friends and Family shortly after the November election. "Look around," this Friend said. I turned my head to scan the panoramic views of Manhattan, the winding marble staircase, the original Monet on the walls, the untouched crystal plate of macaroons on the table. "I'm not a loser. Hillary is a L-O-S-E-R," the Friend said, making an *L* with one hand and holding it against the forehead.

the Sunday *Times*. She also didn't want Anthony Weiner's latest dick pic, this time with the pixilated image of his three-year-old son, Jordan, in bed next to him, to appear on the cover of the *New York Post* (POP GOES THE WEINER). Or for her eight-point postconvention lead in national polls to have evaporated to three points. But before Labor Day, Hillary would get all of this—plus a budding case of pneumonia.

44

"Media Blame Pollen"

LABOR DAY 2016

"Hope you're enjoying a last weekend in the Hamptons before things get crazy," Brown Loafers closed his brief but effective email shaming me for the "Hillary and the Ultrarich" story.

> **Westchester County Airport Departure**
> **Date: Monday, September 5, 2016**
> **Media Arrival: 6:00AM EDT**
> **Time: 9:30AM EDT**
> **Where: Westchester County Airport–Ross Aviation**
>   **West FBO–White Plains, NY**

I wished I were in the Hamptons. On the last Sunday of the summer, I was on my way to White Plains, a sturdy commuter town, an hour north of the city, that is the opposite of the Hamptons. We had to be at the airport near Chappaqua early the next morning to cover the most anticipated event of the campaign (at least to us): The unveiling of a squat Boeing 737 strewn with a shiny new coat of white paint, an *H* on its scion-blue tail and the newly minted STRONGER TOGETHER slogan splashed across its side torso. The plane had arrived.

The Travelers welcomed this development with the silly enthusiasm of neglected hamsters about to be handed a carrot. We took selfies in our assigned seats, piled stacks of H-branded cocktail napkins into our bags, and shot Snapchat videos of our shared quarters.

"Look! There's Huma . . ." one of the Travelers said. Mike, the campaign's freckled luggage handler, pulled a sheer navy-blue curtain closed to block off the front cabin where the staff took their seats. We hadn't seen Huma since she announced her separation post–POP GOES THE WEINER.

"I guess that answers whether Hillary's gonna sideline her . . ."

"Dude, she's never gonna sideline her."

"Poor Huma."

After a few rounds of Whole30 and regular workouts on the road with Annie, I'd managed to gain only a few pounds. But I remembered from '08 that the real gluttony came with the in-flight catering.

When we boarded the Stronger Together Express, plates of fresh fruit and cheese, yogurt parfaits and bottles of sparkling water sat on our downturned tray tables. (For all the legends of the inebriated traveling press corps, we preferred guzzling down buckets of Perrier.) We'd already been fed Dunkin' Donuts in the hanger while waiting on a team of German shepherds to sniff our luggage. But when the flight attendants came down the aisles listing the breakfast offerings, I said, "Oh, I just had a donut. I guess I'll take the omelet."

We were so fat and happy on the plane that we adopted a new word— the *slunch*, short for the second lunch. The slunch usually arrived around 3:00 or 4:00 p.m. and often had a local theme. Lobster rolls left on our seats after a stop in Boston. (Annie sat on hers.) Empanadas after a rally in Miami. Barbecue on our flight out of Kansas City.

I couldn't have timed my Hamptons story any worse. The first day we finally shared close quarters and no one from the campaign, not even Barb, the petite photographer whom I'd been friendly with since 2008 and who went on the Africa trip in 2012, was speaking to me. I'd walked into the private-jet terminal in Westchester acting as if it were the first day of school. "Hey, how was everyone's weekend?" I said to a private-airport lounge full of campaign staffers' eat-shit looks.

"I just thought that story was really unfair," a young press aide said as we washed our hands and partook in the free mouthwash in the ladies' room. "What's she supposed to do, *not* fund-raise?"

The story, cowritten with Jonathan Martin, was meant to be a light, airy romp explaining to readers how Hillary spent the last couple of weeks of August. We filled it with vapid Yorkie details. The $100,000-per-couple lamb dinner that Lady Lynn Forester de Rothschild hosted under a tent on the lawn of her oceanfront Martha's Vineyard mansion. The ten-person chat in Sagaponack that raised $2.5 million. Hillary sandwiched between Justin Timberlake and Jessica Biel in a photo booth set up at a $33,400-per-guest lunch at the couple's Hollywood Hills home.

The campaign called it snarky. The #ImWithHer crowd called it a hit job.

I'D SPENT THE summer mostly writing glowing stories, not because I wanted to suck up—by then Hillary's and my relationship, which had early on wavered between courtship and repulsion, had been undone—but because I wanted to give Hillary her rightful front-page (above the fold) due for becoming the first woman to capture the Democratic Party's presidential nomination, standing up to Trump, and pulling off a smooth, even inspiring, convention.

The campaign wasn't perfect. Hillary wasn't perfect. And I got my fair share of hell for covering those flaws in real time. Much of what I got, from Hillary and her supporters, I deserved. But the vast majority of my stories had nothing to do with emails or the Foundation or Hillary's perceived flaws. Taken as a whole, they'd rank neutral to positive, with plenty of wet kisses thrown in.

Ever since the *Journal*, I'd been drawn to policy stories. Before the 2016 campaign even started, I wrote about Hillary working to craft an economic agenda that would address growing inequality and stagnant wages. For the next two years, I wrote about every one of Hillary's major policy rollouts—taxes, infrastructure, immigration, gun control, early childhood education, and others—all the while knowing that these stories would go virtually unnoticed by my editors, by readers, by the campaign.

And I gravitated to lengthy features about little-known chapters of Hillary's life. This was not easy to do with a presidential candidate who'd first been featured in the *Times* in 1969, five years before the paper first published the name Bill Clinton, and who remained in the public eye ever since.

Tension didn't always mean negative stories. I wrote features about Hillary's tortured relationship with her sullen, unappeasable father, Hugh Rodham; about the two years after Bill Clinton lost reelection in the 1980 Arkansas governor's race, when it fell on Hillary to find a place for the family to live, care for nine-month-old Chelsea, and pick up extra hours at the Rose Law Firm, all while Bill cheated and sulked, playing "I Don't Know Whether to Kill Myself or Go Bowling" on the jukebox. I wrote about Hillary as a fish-out-of-water litigator trying cases in Arkansas, including defending a pawnshop bouncer named Tiny (because he was anything but) who stood accused of beating up his girlfriend and a crop duster who flew his plane too close to the fields and injured a farmworker.

I thought of this as explanatory journalism and that my features would help Americans see this enigma of a public figure as a daughter, a young activist, a jobbing lawyer, and a working mother stressed about her family's finances.

These pieces required digging through documents in Arkansas, a welcome change from the pack journalism of the bus, and almost always humanized Hillary in ways her campaign often failed to do, yet The Guys either ignored my interview requests on these stories or tried to kill them outright. Hillary (whose friends always said she was "the most famous person no one knows") didn't want to leave anything to chance. If she was going to reveal glimpses of her biography, it would be on her own terms.

In the end, none of it mattered. I'd always tried to see and cover Hillary as a complete person, with black and white and lots of gray areas, but there was never any gray area in how Hillary saw me. No number of positive, front-page stories could change her mind. And I understood why Hillary hated the Hamptons story in particular. Mingling with the .001 percent looked terrible. But it was true, all of it.

WE'D HARDLY LAID eyes on Hillary when we all unbuckled our seat belts and poured into the aisle to watch her stride to the back of the plane. She walked past the rows of Secret Service and staff to greet our rowdy rear cabin. Hillary was Labor Day casual, in the cotton baby-blue shirt with a wide collar like the one she'd worn to the Iowa State Fair. Brooklyn proposed Hillary have a barbecue with the Travelers and our families before the inaugural flight, an easy way to butter us up before the general election. The idea never got past Huma. "Not happening," she said.

"Hey Guys. Welcome to our BIG PLANE. It's *so* exciting," Hillary said in a tone that communicated the excitement of a visit to the post office. "I think it's pretty cool, don't you?" No one said a word.

"You're supposed to say yes," Hillary said.

She surveyed our blank stares and tried again. "I am sooooooo happy to have all of you with me," she said, clutching her fists and stomping them up and down her torso like a toddler demanding a toy. "I've been just waiting for this moment. No, really, and I'll come back to talk to you more formally but I wanted to welcome you on to the plane."

I thought of the time in 2008 when Hillary stepped onto the bus in Des Moines bearing bagels and coffee that no one accepted. As much as I'd whined and pushed for a shared plane, by the time the Stronger Together Express arrived, we got the message. Hillary didn't want to finally get to know the women (plus Dan Merica) of her press corps. There would be no late nights gossiping off the record over a goblet of cheap wine. But with two months left, facing an opponent who ruled the airwaves, Hillary decided she needed us.

Later that afternoon Hillary assumed the position of a QVC host walking back to press quarters to display at chest height a copy of her new *Stronger Together* book, a collection of Clinton-Kaine policy proposals on sale in your local bookstores starting Tuesday for the bargain price of $15.99. Trump's 757 sat on the tarmac in Cleveland in a kind of high-noon showdown with the peppy Stronger Together Express. "I heard now that we've got this great plane that Donald Trump actually invited his press on his plane where I'm told he even answered a few questions," Hillary said. "So following my lead as he just did I would hope he continues and releases his tax returns."

Part of a campaign reporter's job is allowing yourself to be used. At worst, we are captive stenographers, the Tripods at the back of the gym. At best, we are the unruly conduits to the American people. The campaign must endure and accommodate us to get its daily message out. But I'd never felt more like a *NY Times* Presstitute than on the inaugural Stronger Together flight.

I started to think of myself as a secretary, valued by my employer mostly because of my ability to crank out 120 WPM on my lap in a speeding motorcade. I'd record the audio of Hillary's rallies or press conferences and, as we piled into the motorcade to get back on the plane and fly to the next swing-state city, transcribe the quotes that fit the day's theme (i.e., Trump's taxes, Trump's treatment of women, Trump's bankruptcies). I'd type so fast that my cursor had a hard time keeping up and then send a feed on to one of my colleagues in New York, who wrote the main "lead all" Frankenstory that day, which was almost always Trump focused.

Or I'd pull any Hillary news into a tight eight-hundred-word "daily," the obligatory news stories that typically ran online or in the vitamin pages. I could write dailies half-asleep while drinking red wine out of a plastic cup.

Six hundred ten words, web only:

> Moline, Ill.—Hillary Clinton accused Russian intelligence of interfering with the American election, implying that President Vladimir V. Putin viewed a victory by Donald J. Trump as a destabilizing event that would weaken the United States and buttress Russian interests . . .

Back in the newsroom, my *Times* colleagues made calls, pored over documents, and even got Trump's taxes in the mail. But the Travelers, well, we were the yeomen of political journalists, shuttling between states to hear the same speech over and over and fighting over Gogo in-flight wireless that was so temperamental that we'd started calling it NoGo.

But even on days when the K-9 crew that sniffed our luggage logged a more productive day's work than I did, I believed the trail was about more than seeing the country (or at least its high school gyms), being immortalized in a pile of front-page bylines, and having some good stories to tell the kids (that was, if my fertile window hadn't closed). I was never one of those political reporters who, reared on a diet of *All the President's Men*, held lofty notions about holding the powerful accountable and uncovering a scandal that leads to the toppling of an administration, international fame, and getting paid to blab on cable TV in perpetuity. That wasn't me. I just wanted to tell good stories that helped explain the world to people.

In Japan, I started to see myself as a cultural anthropologist, using my reporting to demystify a culture that many American readers had reduced to a few geisha flicks and the atomic bomb. I had the same self-important notion when it came to Hillary.

In minutes, and over Katy Perry lyrics blared from scratchy speakers, I could weave into my stories references to Hillary's winning campaign for student government president at Wellesley (which the college paper called "vague as to exactly how" she "would implement the change in the power structure"); her work on the Children's Health Insurance Program as first lady; her stance as secretary of state that the 2011 Russian parliamentary elections had been rigged, important backstory for the argument she was now trying to make about Russian meddling in the US election. I could even drop references to the first time Hillary ordered room service—she was a Goldwater girl staying at the Fontainebleau hotel on Miami Beach to see Richard Nixon nominated at the 1968 Republican National Convention.

I hoped this context, my up-close observations of Hillary, her aides, her supporters, her reception in black churches and inner cities and steel towns that she'd zip in and out of, would help readers feel like they better understood this unknowable being. That they could walk into the voting booth on Election Day a little more informed about their choice. If we really do all have a part to play in this messy, mesmerizing process that every four years alters the course of the country, then my mental trove of Hillary trivia and hammering out words on a rickety tray table on a refurbished 737 was mine.

THE COUGH STARTED in Luke Easter Park on Kinsman Road in a black area of Cleveland.

"Hey Cleveland! Happy, happy Labor Day," Hillary said, letting out the first phlegmy whopper. "When we were trying to figure out where we could be, we all said, 'Let's go to Cleveland!'"

Her voice cracked. She coughed again, five, six, seven, eight times. She reached under the podium and took a sip of water. Campaign staffers winced. Voters looked at each other with clenched jaws. Somebody started a lukewarm chant of Hill-a-REE, but Hillary, now in a full coughing fit, put her hand up to silence the crowd. She convulsed and patted her chest with her hands a few times. A prayer for this to pass.

"I've been talking so much . . ." she started, but she couldn't finish.

Thing was, Hillary hadn't been talking that much. She'd spent the last couple of weeks talking remarkably little. She unwrapped a honey-flavored Halls and managed to joke, "Every time I think about Trump I get allergic." But it came out more like an extended wheeze.

The coughing attack pained me. Yes, Hillary was extremely pissed at me that day, but she didn't deserve this. She didn't deserve to spend sixteen months on the road only to play into all the conspiracies about her health. The *Drudge Report* led with a photo of Hillary's head cocked back mid-cough with the headline CHOKE! MORE PAIN ON PLANE and eight accompanying stories including "10 Doctors Question Hillary Health," "Media Blame Pollen," and the requisite "Complete Timeline of 2016 Coughing Fits."

The next day, Hillary came to the back of the plane to tell us that twice a year, in the spring and the fall when the pollen comes out, her allergies flare up. "I just upped my antihistamine to try to break through it," she said. "It lasts a couple days and then it disappears."

Drudge was right. I wanted to blame it on pollen. I wrote a lighthearted blog post about Hillary's allergies in which I compared conservative media's innuendo about her health to conspiracies about Obama's birth certificate. I quoted Hillary telling us, "I have created so many jobs in the sort of conspiracy-theory machine factory. Because honestly, they never quit."

But Hillary was clearly sick. Four days after the Labor Day croup, she clutched onto the lectern at the New-York Historical Society and said in a frail, diminutive voice, "I believe that America's national security must be the top priority for our next president . . ."

I was sitting in the front row, a few feet away. By then I knew it wasn't pollen. Hillary had drowsy eyes and the reduced presence of a person who had popped a handful of extra-strength Tylenol when they should be in bed.

Three hours later, at an "LGBT for Hillary" fund-raiser in the cavernous ballroom of Cipriani in lower Manhattan, Barbra Streisand sang a Trump-themed parody of "Send in the Clowns." ("Is he that rich? / Maybe he's poor? / 'Til he reveals his returns / Who can be sure? / Who needs this clown?") Laverne Cox introduced Hillary, and she walked onstage in front of the blue velvet curtain to a standing ovation. She acted loopy from the start.

"Wow. Thank you. Thank you. It's sort of like the seventh-inning stretch," Hillary said, resting her hands on the clear Lucite podium. "You know, I've been saying at events like this lately, I am all that stands between you and the apocalypse. Tonight, I'm all that stands between a much better outcome!"

It got worse.

Forgetting that the campaign allowed the press pool to cover her remarks, Hillary did her usual bit about the baskets. But like the 3-D printer that, the more times she said it, moved in Hillary's mind from Waterloo to Moline, the three baskets of Trump voters had somehow become two. "You know, just to be grossly generalistic, you could put half of Trump's supporters into what I call the basket of deplorables," Hillary said, pausing for comedic emphasis. "Right?"

Amirite, folks?

Babs was said to have cringed backstage. The Guys tried the "Clean Up in Aisle 7" approach. They explained that Hillary's comments weren't any different from what she'd been saying all along about the racism Trump's campaign has fueled.

By 2:06 p.m. the next day, Hillary issued a statement saying she regretted that she'd said "half." Podesta released his own statement, saying Trump "has spent 15 months insulting nearly every group in America" and that "this is without a doubt deplorable—but this is who he is."

After the convention, donors asked Brooklyn what they planned to do to pull Hillary's trust numbers out of the toilet. The answer was always the same: nothing. Podesta would explain, "I remember when no one trusted Bill Clinton, and he won twice." That was true, but "Slick Willy" was different. Voters may not have trusted Clinton with their daughters, but they trusted that he was looking out for them. That wasn't the case with Hillary, and laughing at tens of millions of Americans while surrounded by her rich, fabulous urbane friends at a fund-raiser at Cipriani wasn't going to help. "I really messed up," she told aides that night.

ON SUNDAY MORNING, Hillary arrived at Ground Zero to commemorate the fifteenth anniversary of the September 11 terrorist attacks, and I dragged Bobby to a Flywheel class. Despite his cultural aversion to healthy food and light beer, Bobby is slim with broad shoulders and legs so long they stretch the length of our L-shaped sofa. I didn't usually have the time or energy to do much marital nagging, but with the election in sight, I had visions about the active weekend routines that we'd soon adopt. With Hillary at the September 11 memorial service and the press pool keeping an eye on her, I had a free morning to lure my husband to a spin class.

"C'mon, it will be fun," I said the night before. I sat on the chaise, bought at a clearance sale, that overlooked Seward Park and that I hardly ever had a chance to lay in.

Lately, whenever I looked out at the trees and the squirrels, I thought of Hillary. It wasn't enough that we lived on Clinton Street, but our second-floor apartment overlooked Seward Park. Hillary had been telling donors that William Seward was her favorite secretary of state. "He was, after all, a senator from New York who ran for president, lost to an eloquent up-and-coming politician from Illinois and then surprised everyone by serving as his former rival's secretary of state . . ."

Bobby sat on the sofa.

"Can't we just go to brunch?" he said.

I knew Bobby didn't want to spin when he proposed brunch, a New York activity that for him ranks right above dragging an AC unit home from PC Richard.

"We can go to brunch after!" I said.

"I've never been spinning before," Bobby said.

"It's so easy. I'll show you how to set up the bike and everything."

"Okay," he said, without turning around. "But I want to be in the back row."

We'd rented our cycling shoes and were heading into the studio when I looked at Twitter one last time before tuning out for a whole forty-five minutes. "Oh my God," I said. "Oh my God. Oh my God. Oh my God." People were staring. I'd stopped in the doorway and had caused a pileup of girls in sports bras.

"What is it?" Bobby said, scooting me over a few inches. Our cycling shoes clinked.

"Hillary passed out. She passed out. I've got to go. I'm sorry," I said. "You'll be great."

I kissed him and unvelcroed my shoes without sitting down. As I raced out of the studio, the redhead behind the counter, who heard my breakdown, yelled, "We've got some zinc pills here if your friend Hillary needs them."

"No thanks," I said. "She's good!"

But my friend Hillary was anything but good. I sat on a bench at Astor Place and watched and rewatched the video a passerby shot of Hillary's legs giving out, her limp body being hoisted into her Scooby van. It looked like one of those "Hillary's health" hoax videos assembled in Matt Drudge's basement. I ran the footage by Hired Gun Guy to make sure it wasn't fake. Half of me hoped that it was.

Brown Loafers, after not confirming whether Hillary had even left the memorial, put out a statement saying she "felt overheated so departed to go to her daughter's apartment and is feeling much better." It wasn't his fault. Hillary only told Huma and Cheryl about the pneumonia diagnosis she got on Friday.

Hillary was on her way to Chelsea's apartment in the Flatiron District when the unsolicited texts from sources started flooding in.

> #1: Where does the person go who has been shot in the robbery? Never to the hospital.
> #2: This is a disaster.
> #3: She said she was fine and was hiding her true health.
> #4: Heard SS made her wear heavy body armor . . .

. . . and, later that afternoon from my mom:

> #5: I hope Hillary didn't give you pneumonia!

David Axelrod tweeted the primo question: "Antibiotics can take care of pneumonia. What's the cure for an unhealthy penchant for privacy that repeatedly creates unnecessary problems?"

Bobby walked out of the class like he'd spent the day horseback riding, his shirt glued to his chest.

"Hi, how was it? I'm really sorry, but I'm on deadline now . . ."

He sat on the bench next to me and took a long swig of water without speaking.

"The good news is brunch is out," I said.

He took a couple more sips and, squinting into the traffic, said, "How's Hillary?"

$$45$$

The Fall of Magical Thinking

ON BOARD THE CAMPAIGN PLANE, SEPTEMBER 2016

The clementine crushed any hopes we had that the Stronger Together plane would lead to Hillary wandering back to schmooze with us.

After a couple of unsuccessful tries, Ruby Cramer rolled the fruit up the aisle and into the front cabin. We'd written on its peel the question, "Would you rather have dinner with Trump or Putin?"

The Travelers waited anxiously until the clementine came whirling back down with Putin circled. We tweeted out Hillary's response. That's when Brown Loafers came back and informed us the clementine had been "off the record." He said he'd circled Putin, not Hillary. The debate went like this . . .

"But you never said the orange was off the record!"

"It's a clementine," Brown Loafers said.

"You don't have a case!"

"Everyone is really tense," Brown Loafers said, shaking his head.

The campaign won. The embeds tweeted out a clarification: "HRC saw the orange, noted she once dined w Putin, but did not issue an answer either way."*

On one stop, Hillary came to the back of the plane to prove how much she appreciated her young staffers. "How's Arun treating you? Arun goes all the way back to the State Department with me, don't you? How many years has it been, Arun?" Hillary said singling out our

---

* It was not lost on the more senior reporters that Ruby Cramer, whose father, Richard Ben Cramer, had revolutionized access journalism and been so close with sources that Biden spoke at his memorial service in 2013, was rolling an orange (sorry, a clementine) up the aisles trying to get something—anything—out of Hillary.

blushing, bushy-haired press wrangler who almost always wore a jacket and tie. We were all too polite to tell her his name was Varun.

Hillary popped into the press cabin, freshly shellacked and ready to get the televised "I'm feeling GREAT" exchange over with.

"Hi, guys! Welcome back to Stronger Together," she said.

Andrea Mitchell took the first stab: "How are you doing? How are you feeling?"

"I am doing GREAT, thank you so much," Hillary said.

Imagine two cars stuck in third gear. Each is being pushed up opposite sides of a hill, but just before the drivers get to the top and can see each other, they keep sliding back down the hill. That's what it's like trying to make on-the-record small talk with Hillary.

She promised us she'd take our questions after her speech in Greensboro, and when she started to turn away, a Traveler yelled, "What have you been doing for the last few days?"

"Uh, um. I will talk about that later, too," Hillary said, still on the move.

I said the first thing that came to mind: "Did you binge-watch *The Good Wife*?"

"It's done. I am so sad. I really am. It's really a loss," Hillary said. She stopped and turned, lingering over the flash of cameras. "*Madam Secretary*, however, is coming back, so that's something to look forward to."

"Is that odd to watch? It's so meta," I asked.

"I actually get a big kick out of it," Hillary said.

"Yeah, I guess seeing Téa Leoni play you . . ." I said.

"I watched it with a little bit of skepticism at first. But I got so into it. I really like the story lines. They have some good quasi-realistic story lines, so anyway . . ."

"WHY NORTH CAROLINA?" a TV correspondent yelled.

"Excited to get to North Carolina!" Hillary said. She gave a clumsy half thumbs-up and headed to the front cabin.

"Bye . . ." a couple of us muttered.

LOOKING OUT THE window from the motorcade at the relaxed, abundantly green blur of Greensboro, North Carolina, I wanted to spend more than an hour there. But the days when we actually got to know a place ended a long time ago. The campaign was surgical at this point. Get in at this church in Durham or that community college in south Florida and get out.

Hillary walked onto the stage to James Brown's "I Got You (I Feel Good)" with such aplomb that conspiracy theorists thought she had a body double. At the press conference following the rally, it was Hillary all right.

I sat cross-legged on the wooden floor of the basketball court half listening. My deadline was approaching, and my editor emailed me, "I need a top . . ." which meant I had to at least get two hundred words to him. Hillary agreed to take questions mostly to prove how much "stamina" she had, but as soon as she exhaled and opened her three-ring folder on the podium in front of us, she seemed only half herself. She'd told the rally that the three days at home resting had been a "gift" that allowed her to "reconnect with what this whole campaign is about."

This reassured me since I still didn't know what the campaign was about. The race was pretty much tied at this point. Outside, a cluster of protesters held handmade posters with the usual HILLARY FOR PRISON, plus a newer rallying cry, I AM DEPLORABLE.

Hillary spoke softly and with minimal animation. "I want to give Americans something to vote for, not just against," she said. "I want to close my campaign focused on opportunities for kids and fairness for families. That's been the cause of my life. It will be the passion of my presidency."

That sounded good, but Hillary must've still been on heavy medication if she thought she could make the rest of an election against Trump about the "detailed plans in thirty-eight different policy areas" that she'd laid out over the past year and a half. Instead, she would spend much of the fall deriding Trump for his fat shaming of a former Miss Universe and defending Rosie O'Donnell ("an accomplished actor") and Kim Kardashian against Trump's "pathetic" put-downs.

Hillary got irritated with us even faster than usual.

When the Travelers asked several times about why she hadn't told Tim Kaine about her pneumonia diagnosis on Friday, and "what does that say about what your relationship would be like with him in the White House?" she said, "We communicated. We communicated, but I'm not going to go into our personal conversations."

Asked why it took so long for the press and the public to know, Hillary blamed Brooklyn. "My campaign has said that they could've been faster, and I agree with that. I certainly expect them to be as focused and quick as possible." The only problem with that excuse, of course, was that her campaign didn't know.

CHUCK SCHUMER, WHO'D been next to Hillary at the September 11 memorial service, later said he'd had pneumonia, too. The virus became a status symbol of physical proximity to the next president. Carolyn joked, "Did you see that Jennifer Granholm and Chris Quinn infected themselves with pneumonia bacteria so they can announce they have it like Schumer did?"

Like most of the press, I considered Hillary's handling of her pneumonia and what it said about her psychology more of a story than any actual health problems. She was sixty-eight, and we could hardly keep up with her ourselves. But going back to the *Guernica* press conference, the biggest fuckups of Hillary's campaign were always because of her time-warp syndrome. Hillary had the 1990s in mind when she told only Cheryl and Huma about her pneumonia diagnosis.

After the September 11 memorial, in the van on the way to Chelsea's apartment, Huma said to Hillary, "How does the statement about dehydration square with the pneumonia diagnosis?" The rest of the aides in the van looked at each other. *Uh, what pneumonia diagnosis?*

During the 2008 primary, whenever critics would accuse Hillary of living in the past, she'd say, "What part of the 1990s didn't they like— the peace or the prosperity?" She didn't talk about the '90s much in 2016, but everything, from her rallies to TV ads and piles of opposition research and how she handled the pneumonia, came out of the 1992 playbook. Cheryl and Huma, who were with her in the White House, appreciated her "zone of privacy." They'd lived through Whitewater and Travelgate and all the other *gates* James Carville loved to recite on cable news. Disclosure didn't ever help matters.

Brooklyn often suggested that Hillary try to break out of the '90s mind-set. Play to her strengths and cut back on the traditional rallies. Her pint-size crowds compared to Trump's only fed reporters' reductionist theories about Enthusiasm anyway. She could do town halls and impromptu stops, aides suggested. Like the time she walked into a Girl Scouts office in Ashland, Kentucky. "I loved being a Girl Scout. I was a Brownie first, then I was a Girl Scout, then I was something called a Mariner Scout when I was in high school," she told the little girls who gathered around in their green and brown patch-covered sashes.

Hillary seemed more at ease, less scripted in these settings, and they always got picked up on the news the same way a rally or speech would, Brooklyn argued, but Huma always insisted that they keep it old school: big (or biggish) rallies, TV ads, and, above all else, debate prep.

Hillary even rehearsed how to create a prime-time TV moment, like the one her husband had in a 1992 debate when George H. W. Bush checked his watch as Bill had a heartfelt exchange with a woman about the down economy. Hillary would take a couple of steps toward the questioner, make eye contact, empathize. Again, Hillary cut back on her campaign schedule and spent most days preparing for the debate in a conference room at the Doral Arrowwood Resort in Rye Brook.

Thursday, 9/22—HRC has no public events scheduled.

Friday, 9/23—HRC has no public events scheduled.

Saturday, 9/24—HRC will attend the Opening of the National Museum of African American History and Culture in Washington, DC.

Sunday, 9/25—HRC has no public events scheduled.

Monday, 9/26—HRC will participate in the Presidential Debate at Hofstra University in Hempstead, NY.

I'd been at the office the day Hillary came to the belly of the beast, the New York Times Building, to meet with the editorial board. After about an hour of policy talk, Carolyn tried to show Hillary the *Times'* latest technological innovation, a 360-degree video camera. A couple of weeks earlier, after Trump met with the editorial board (when he said his proposed border wall was "negotiable"), he'd delighted in the funky little multilens camera, peeking into its mirrored eye and asking how it works.

Hillary looked at Carolyn as if she'd tried to hand her a ticking blob of C-4. She took a couple of steps back and called out for her team. "Uh, um, Jen! Jen! Can you take a look at this?" Then Hillary turned to Carolyn, the editor who oversees the *Times'* political coverage, and said, "You need to talk to my team about that."

Hillary always used debate-prep sessions as cathartic exercises. Before the "you're likable enough" debate in New Hampshire in 2008, Hillary had lost it. She kicked everyone out of the room and sipped hot tea with Chelsea. Aides understood that in order to keep it all together onstage, Hillary sometimes needed to unleash on them in private. "You want authentic, here it is!" she'd yelled in one 2016 prep session, followed by a fuck-laced fusillade about what a "disgusting" human being Trump was and how he didn't deserve to even be in the arena.

The *Times* devoted at least a week's worth of energy into trying to figure out who would play Trump in debate prep. My unsubstantiated guesses had included Andrew Cuomo ("It has to be someone Hillary genuinely hates," I'd argued) or Terry McAuliffe. I felt like an idiot when Maggie Haberman broke that they'd tapped Original Guy, who later told Annie Karni that he got off his meds for the occasion. *Hmmm, wherever will Hillary find a manipulative, sometimes-charming, often hilarious, possible sociopath?* Hillary also relied on Ron Klain, a Biden loyalist who had prepped Bill for the I-feel-your-pain moment in 1992.

When Pat Healy interviewed Trump around that time and ran some of Hillary's debate-prep techniques by him, Trump responded, "I don't need to rehearse being human."

46

Debate Hillary

LONG ISLAND, NEW YORK, SEPTEMBER 26, 2016

The morning of the first general-election debate, I slept until eight. I did a hip-hop yoga class in a candlelit, one-hundred-degree room and then sat in Seward Park drinking the first of a half dozen iced coffees I'd ingest over the next twenty hours. Watching the elderly Chinese ladies doing tai chi in Seward Park always soothed me. I wanted to join in on their languid arm circles and ninja-like toe pivots, but I worried that I didn't know the moves.

A few weeks earlier I'd asked my friend and neighbor Martin Wilson, "How do they all know exactly what to do?"

"It's like the hokey pokey for them," he said. "They just do."

Debate day was kind of like that. We had our own rhythms. It starts out quiet. Print reporters typically sleep in, work out, and indulge in a carb-heavy breakfast. By midday we are gaming out different scenarios and assembling a thousand words of "B matter," the paragraphs of generic background that we can cut and paste into a story on deadline. ("A vast, plodding political machine surrounds her, insulates her and protects her. But Mrs. Clinton is at her most compelling—and sympathetic—when she is all alone, defending herself in these tense, hard-to-watch moments.") Then comes an extended anticipatory period between 4:00 p.m. and the start of the debate, 8:00 or 9:00 p.m. This is when pizza or sub sandwiches arrive in the newsroom and we reach our limit of listening to cable TV pundits. Someone picks up the remote and yells, "Mind if I mute this?"

As soon as the candidates walk on the stage, we do not speak. For the next ninety minutes there is only the occasional yell of "Did anybody get that line?" and "Can someone send me that exchange?"

I arrived at Hofstra University nine hours before the debate started, which seemed a reasonable amount of time to check in and make sure the Wi-Fi worked. I found Hired Gun Guy looking telegenic in heavy foundation after an MSNBC hit.

Reporters pounced. In my continued misreading of the scrum, I told him my hot-yoga predebate ritual and asked about Hillary's. (I didn't know Hillary's debate-day rituals, but I did know something about her workouts. She'd said she does yoga "my own way." I once asked if this type of yoga incorporated briefing books. A girlfriend talked her into trying P90X home videos, and Bill told me the previous summer, "Hillary has been working out—regimen in the morning six a.m. in the pool. I thought it was easy and she made me try it with her.")

"Did Hillary walk her dogs? Did she do yoga? Do she and Bill binge on blueberry pancakes as a debate-morning tradition?" I asked.

"We're not that kind of campaign," Hired Gun Guy said.

HILLARY KNEW THIS was the biggest night of her career. "Somebody said to me, 'Well, remember there'll be a lot of people watching—a hundred million people watching. And sixty million will be paying attention to the campaign for the first time,'" Hillary said at a fund-raiser in East Hampton. (The press had been kept in a basement TV room and could hear her remarks through the ceiling.)

Hillary was always so insecure after debates. Almost a year earlier, after the first primary debate in Vegas, when Sanders had declared that "the American people are sick and tired of hearing about your damn emails," Hillary had appeared so masterfully competent that she'd crushed Biden's lifelong dream in two televised hours. She went backstage at the Wynn and asked her team, "How'd I do? Was I okay?" The Guys erupted, reassuring her all at once. "Are you kidding? That was unbelievable."

There was no real reason for Hillary to doubt herself. Aside from the time in 2007 when she gave a yes-no-maybe response to whether she supported giving driver's licenses to undocumented immigrants, she almost always killed it.

The debate is to Hillary what the handshake is to her husband, what the soaring oratory is to Obama. This preternatural gift went back to high school. A childhood friend, Ernie "Ricky" Ricketts, told me about the time Hillary, such a die-hard Goldwater girl that she wore a hat with the AuH_2O logo, had to play LBJ in a mock debate of the 1964 presidential election. She protested but eventually pored over Johnson's civil rights, foreign policy, and health-care positions and presented her case with such ardor that by college, she'd even convinced herself and became a Democrat. "Whenever we would say, 'Well, who could speak up for this or that for us?' It was always, 'Well, let's get Hillary to do that,'" Ricketts recalled.

I'd covered her in more than thirty of them, and each time I watched her onstage, I felt something almost like pride. In 2007, she'd shown her biting wit, including joking that she was wearing an "asbestos" pantsuit. She mastered the art of playing the victim. ("Maybe we should ask Barack if he's comfortable and needs another pillow," she said in another 2008 primary debate, after *Saturday Night Live* portrayed a coddled Obama in a debate skit.) In New Hampshire, she'd earned the sympathy of women voters after the moderator asked about her likability problem. "Well, that hurts my feelings . . . but I'll try to go on," leading to Obama's "You're likable enough" quip.

All of those debates seemed so inconsequential compared to this one. The first woman nominee on the general-election debate stage against a reality TV star renowned for misogyny.

Karen Dunn, the petite lawyer and preeminent Democratic debate coach, gave Hillary a fist bump before she went onstage. She reminded her to smile, since viewers would see a split screen while Trump spoke. "I got this," Hillary said.

SHE WAS RIGHT. Hillary won the debate. Hillary won all three of the debates. After the final debate in Vegas, OG hugged Hillary and told her she was a "badass hombre."

I won't bore you with all the details because in the end the debates didn't matter. Our rituals, the *Times*' and Hillary's, our antiquated notion that voters would assess the candidates as we did, failed us. Competence, preparedness, policy. These were words the privileged used. Turned out a lot of people just wanted to blow shit up.

The irony was that after the debate, Hillary, for the first time in months, had been cocky. There was none of her usual self-doubt and *how'd-I-do*s. She hugged aides and donor friends, financiers mostly. She told them that they didn't need to worry anymore. "I will win this," she said.

She got the Republican nominee to admit he hadn't paid federal income taxes ("That makes me smart."), shimmied her shoulders ("Whoo! Okay!") in a five-second burst of humanity that went viral, and dispensed the trove of opposition research she'd memorized with the sweet precision of a PEZ dispenser.

But her favorite part of the debate—the moment when she knew she had him—came at the very end, when Trump was asked why he said Hillary didn't have a "presidential look."

Trump squirmed out of it, pivoting to her lack of "stamina." Hillary saw her opening.

"One of the worst things he said was about a woman in a beauty contest. He loves beauty contests, supporting them and hanging around them," she said. "And he called this woman 'Miss Piggy.' Then he called her 'Miss Housekeeping' because she was Latina. Donald, she has a name—"

"Where did you find this? Where did you find this?" Trump said.

"Her name is Alicia Machado."

Twenty-four hours later, her name was everywhere.

Trump spent the next several days fat shaming Machado. "She gained a massive amount of weight and it was a real problem," he told Fox News the morning after the debate.

Brooklyn arranged a press conference call with Machado. They released a string of bilingual Machado-themed ads with footage of Trump dragging the curvaceous Venezuelan to the gym ("This is someone who likes to eat") and sent out guidance so Democrats could blanket the airwaves with Machado talking points.

We all agreed with Hillary that Trump must've been "unhinged" to keep the fight with Machado going. He was self-destructing. We'd taken his earlier declaration, "I could stand in the middle of Fifth Avenue and shoot somebody, and I wouldn't lose voters," as a cry for help. But it hadn't been that at all.

Each day that Trump handed Hillary ready-made attack ads too delicious for any candidate, especially one focused on women and Latinos, to pass up was another day Hillary didn't talk about jobs or health care or debt-free college. Hillary dropped her promise to give voters something to vote for and not just against. Instead of ending her campaign talking about "opportunities for kids and fairness for families," she campaigned with Machado, tweeted about Machado, talked about her at rallies ("I mean, really, who gets up at three in the morning to engage in a Twitter attack against a former Miss Universe?").

Hillary had consulted over two hundred academics and economists to help put together her campaign's policy proposals. A year earlier, at the OTR drinks in New Hampshire, Hillary had berated our pea-size political brains for being uninterested in policy. Now, Trump had made her as devoid of substance as he was.

By the fifth day of Machado-themed programming, the Travelers would say, "Donald, she has a name . . ." and someone would reply, "Her name is Miss Swing State Latina."

47

How I Became an Unwitting Agent of Russian Intelligence

You could define the *beginning* of the end as the point when the protagonist has to see that her actions mean something and that if they don't work out right, she is well and truly fucked.
—JULIE POWELL, *JULIE AND JULIA*

NEW YORK CITY, OCTOBER 2016

The moment you realize that without even knowing it, you've ended up on the wrong side of history. It's dizzying. Months after the election, every time you hear the words "Russia" and "collude," this realization will swirl in your head, enveloping everything. It will be the worst at night. You weren't the only one, but it will sting you the most. And the strange thing is, it started just like any other day.

Friday, October 7, 2016, at 12:15 p.m. Hurricane Matthew barreled up the western Atlantic toward Florida, Georgia, and the Carolinas. Hillary took a break from debate prep to "discuss the storm's trajectory" with the Department of Homeland Security secretary and FEMA. That's what presidents do, and Hillary thought that if people could see what kind of president she would be, they'd vote for her.

An hour earlier, the *Washington Post*'s David Fahrenthold picked up the phone. The voice on the other end asked if he'd be interested in taking a look at some previously unearthed footage of Trump in a 2005 *Access Hollywood* video.

The *Times* newsroom had been quiet that afternoon. I'd fed a statement from the campaign (on background, from an aide) to the main story about the hurricane. "[She] noted her commitment to ensuring that FEMA has the tools and resources it needs to both prepare for and respond . . ."

I needed to get out by 7:00 p.m. to meet Bobby for dinner and the Jason Isbell concert. My birthday had been a few weeks back. Bobby, after a couple of rocky years of odd gifts (a pair of platform sandals he bought at the local shoe repair), had finally adopted my family's opulent observation of every birthday. On any other year, the tickets would've been the perfect gift—I adored the Alabama singer-songwriter. But when I opened the card, I anticipated marital calamity. I thanked him, fighting the urge to yell, "How could you buy me tickets for something A MONTH BEFORE THE ELECTION?" I prayed breaking news wouldn't get in the way.

The *Post* story hit at 4:15 p.m. I heard "Oh my God," and "Oh God," and "Jesus Christ" float from cubicle to cubicle until my largely agnostic newsroom sounded like a Sunday church choir. We were so paralyzed in disbelief watching the footage of the GOP nominee bragging about sexually assaulting women ("I don't even wait. And when you're a star, they let you do it, you can do anything," Trump told Billy Bush. "Grab 'em by the pussy.") that it took several minutes to set in that the *Post* had scooped us bigly.

This is how men and women are different: The men in the newsroom popped out of their chairs. They discussed the implications of the video. They started scribbling skedlines—brief descriptions of upcoming stories—on Carolyn's whiteboard. A story on the quick rebuke by the GOP Establishment. ("Is somebody calling Paul Ryan?") What impact Pussygate would have on women voters. ("He just lost more than half the electorate, if he hadn't already lost them.") The pressure on Trump to drop out. ("There's no way he can survive this.")

I stared into my screen, as frozen as the paused image of Trump and Bush stepping off the *Access Hollywood* bus, the unknowing actress in the fuchsia halter dress waiting to greet them. I thought of Hands Across America rubbing us down in Iowa. I thought of the senior Hillary for America field organizer in Brooklyn who'd harassed his assistant until she begged to be transferred to Colorado. Nobody said anything because Senior Field Organizer was a member of the "Mook Mafia." *And these things happen on campaigns . . .*

I was still in this haze at 4:32 p.m. when WikiLeaks tweeted, "RE-LEASE: The Podesta Emails," along with a link to emails from Podesta's Gmail account, including excerpts of Hillary's Wall Street speeches.

Fuck. 4:32 p.m. I had two and a half hours to read through thousands of pages of emails, try to confirm that they were legit, write a story, and answer my editors' questions. Four hours, if I missed our dinner reservation, and I always missed our dinner reservation. I texted Bobby, "Sorry, some breaking news. Meet you at theater. Can't wait!"

Michael Barbaro and another colleague, Nick Confessore, jumped in to help. The three of us could've pulled off a story on Hillary's paid speeches in time for dinner had it not been for the Russians.

Earlier that afternoon US intelligence agencies said the Russians had been behind the DNC hack. "Earlier today, the US government removed any reasonable doubt that the Kremlin has weaponized WikiLeaks to meddle in our election and benefit Donald Trump's candidacy," the campaign said in a statement. The report confirmed what Brooklyn had been screaming about since Philadelphia—Putin was interfering in the election to hurt Hillary and help Trump. The report didn't say anything about Podesta's emails.

At the same time, Clinton aides, Podesta haters mostly (and there were a lot), cautioned me not to believe the "Russia spin." They said it could've been the Chinese. Or they asked how Podesta, a former White House adviser who had published a study on cybersecurity in 2014, could've used a Gmail account with no two-step verification. Maybe it was, as Trump said, "somebody sitting on their bed that weighs four hundred pounds."

But the campaign insisted the hack had been part of the same "sophisticated misinformation campaign" and warned us that the emails could've been "doctored."

I confirmed that the Wall Street speech excerpts were legit. I'd suspected they were. They sounded like Hillary, not the Hillary I heard on the trail, but the *real* Hillary, the one who charmed donors in the Hamptons and reassured a party in Martha's Vineyard that the financial industry wasn't the enemy. The one who was witty (advising Lloyd Blankfein, the Goldman Sachs CEO, that if he wanted to run for office, he should leave and "start running a soup kitchen somewhere."). She was unencumbered by political correctness (saying the Dodd-Frank financial regulatory bill she lauded on the trail had been passed purely for "political reasons").

I remembered what Bernie asked Pete D'Alessandro when he first hired him in Iowa in 2015—"Do you understand my politics?"—and how I'd struggled to answer that question about Hillary. There in a January 25, 2016, email titled "HRC Paid Speeches," sent by the campaign's research director, Tony Carrk, as "flags" to Podesta et al. and leaked to the world via Julian Assange, was my answer.

In 2014, Hillary said her lifestyle made her "kind of far removed" from the "growing sense of anxiety and even anger in the country over the feeling that the game is rigged." She cited backroom deal making used by Abraham Lincoln and mused on the necessity of having "both a public and private position" on politically sensitive issues. In these speeches, she didn't sound like a "progressive who likes to get things done," but a smart, savvy technocrat at home among the global elite. Hillary lamented, "there is such a bias against people who have led successful and/or complicated lives." She said, "My dream is a hemispheric common market, with open trade and open borders," a comment (in the context of clean energy) that drew a couple more *oh-my-God*s from the politics pod.

We huddled with editors to discuss how to proceed. Hillary's refusal to release the speeches had been such a cause célèbre in the primary that I regularly saw protesters holding signs that said I'D RATHER BE AT HOME READING YOUR GOLDMAN SACHS SPEECHES. Now the juicy excerpts of the most sought-after trove of documents in the election had landed in our laps. But it wasn't a scoop, more like a bank heist.

I never told anyone this, but one time when I'd been visiting the Brooklyn campaign headquarters I found an iPhone in the ladies' room. I wasn't sure, but it seemed to belong to Podesta's assistant because when I picked it up, a flood of calendar alerts for Podesta popped up. I placed it on the sink counter, went into the stall, came out, and washed my hands. I had the sense that someone was watching, like those kids in the famous experiment told not to eat the marshmallows. I left the phone sitting there, worried that if I turned it in, even touched it again, aides would think I had snooped. This seemed a violation that would at best get my invitation to the Brooklyn HQ rescinded and at worst get me booted off the beat for unethical behavior.

I can't explain why in the heat of breaking news, I thought covering Podesta's hacked emails was any different.

The debate in the politics pod—with Carolyn's all-powerful white-board hovering to one side, a camouflaged Sarah Palin bobblehead, a few leftover bags of Fritos, my WENDY FOR TEXAS sign (for irony) displayed on my desk, and the *Access Hollywood* video open on most of our screens—went something like this . . .

"What if burglars had broken into Brooklyn and stolen these off Podesta's desk? Would we cover them?"

"True, these were stolen."

"But it's different. They're all already out there."

"Is it different?"

"I think so?"

"It's not like if we don't write about them, no one will."

"True. They're already all over the place."

". . . and she could've just released the transcripts a long time ago . . ."

I checked the time. 5:37 p.m. I just wanted to write the story and get out of the office. After all those months of controversy, Hillary's Wall Street speeches weren't the main news of the day. (If you haven't picked this up by now, Hillary was rarely the main news of the day.) The desk was chaotic, reporters and editors overwhelmed. We had a Trump–*Access*

Hollywood package that included two stories, one news analysis, a story about GOP reaction, two videos, two annotated transcripts, and a sidebar about Billy Bush. Then there was the hurricane and a separate story by my DC colleagues on the Russian hacking. The desk slotted the speech story, which on any other day would've run on page one, above the fold, at twelve hundred words in the vitamin pages.

Everyone agreed that since the emails were already out there—and of importance to voters—it was the *Times'* job to "confirm" and "contextualize" them. I agreed. I didn't argue that it appeared the emails were stolen by a hostile foreign government that had staged an attack on our electoral system. I didn't push to hold off on publishing them until we could have a less harried discussion. I didn't raise the possibility that we'd become puppets in Putin's shadowy campaign. I chose the byline. I always chose the byline. In some twisted way, I put bylines ahead of my husband, my friends, my inner voice that told me not to look at that cell phone.

We didn't know that for the remaining thirty-one days of the election, WikiLeaks would dump a new batch daily and that we'd spend the rest of the campaign "contextualizing" #PodestaEmails and living in fear that our own incriminating email exchanges would come out in the next pile.

In the end, I cowrote six stories and one blog post off the hacked emails. Two of those stories ran on the front page. Six out of the 1,285 stories I wrote on the Hillary beat for the *Times*. The stories weren't all negative—one of them made Chelsea a heroine who cleaned up her family's foundation (even if Doug Band called her a "brat"), another listed Hillary's veep list. And yet there is no way around it, their existence tainted my entire body of work.

In December, after the election, my colleagues in DC wrote a Pulitzer-winning story about how the Russians had pulled off the perfect hack. I was on the F train on my way to the newsroom. I had no new beat yet and still existed in a kind of postelection haze that took months to lift. I must've read this line fifteen times: "Every major publication, including The Times, published multiple stories citing the DNC and Podesta emails posted by WikiLeaks, becoming a de facto instrument of Russian intelligence."

I'd been called a cunt and a donkey-faced whore and a Hillary shill, but nothing hurt worse than my own colleagues calling me a de facto instrument of Russian intelligence. The worst part was they were right. The *Times* columnist David Leonhardt put it best when he wrote, "the overhyped coverage of the hacked emails was the media's worst mistake in 2016—one sure to be repeated if not properly understood."

I felt bad leaving Michael and Nick at the office that night so I could make it to a concert, but it was almost 8:30 and Bobby texted me from outside the Gramercy Theatre, "Where r u?"

"Go, go, we can finish things here," Michael said.

I sat in the darkened theater. Bobby wore his khaki raincoat, the one he was wearing when I first picked him up in the Irish pub a decade earlier. He gave me the aisle seat. We held hands, but my mind was else-where. I was perpetually elsewhere. For years, I'd listened to a few min-utes of Isbell's rueful lyrics to calm my nerves, at my desk on deadline or while driving Beast from Iowa City to Ottumwa on dark farm roads.

Now Isbell stood, black blazer against a bare-bones stage. I hadn't been in such an intimate venue, so engrossed by acoustic guitar since Austin, when all I wanted was to leave, to move to New York, and to become a journalist. I closed my eyes and let his bluesy vocals fill me . . .

"Everything you built that's all for show goes up in flames, in twenty-four frames . . ."

48

The "Big Ball of Ugly"

OCTOBER 2016

Hillary hardly had time to watch the video when her phone started to ring. Friends and advisers urged her to make calls from Chappaqua. Every faith leader, every "family values" Republican, female executives, activists. Rally them all in shared outrage. But she didn't. How could she? She was married to Bill. Hillary remembered what happened the year before, right around Christmas, when she'd kicked the crazy bear.

Trump remembered, too. He knew his womanizing (or worse) would come out. But he'd effectively turned Bill Clinton into a human shield. Trump dismissed his comments in the video as "locker-room banter," adding, "Bill Clinton has said far worse to me on the golf course—not even close."

In the months since Hillary began her campaign, Roger Ailes was ousted from Fox News after widespread claims of sexual harassment; Bill Cosby was accused of drugging and raping women; Brock Turner, a former Stanford University student found guilty of sexual assault, received a paltry six-month jail sentence; a Columbia University student carried a fifty-pound mattress around to protest the school's handling of her rape allegation. After the *Access Hollywood* video came out, more women came forward saying Trump had sexually harassed them.

I wanted Hillary, the first female nominee for president and a feminist icon, to be the collective voice of a nation on the verge of a cathartic scream. Instead, in those early days after the video hit, and allegations mounted against Trump, Hillary mostly talked about cat videos.

"Now, it makes you want to turn off the news. It makes you want to unplug the Internet. Or just look at cat GIFs," she said at a fund-raiser in San Francisco when the subject of Trump and women came up. "Believe me, I get it. In the last few weeks, I've watched a lot of cats do a lot of weird and interesting things. But we have a job to do, and it'll be good for people and for cats."

Michelle Obama, a first lady untainted by sexual scandal, said what Hillary couldn't. "I can't believe I'm saying that a candidate for president of the United Sates has bragged about sexually assaulting women," Michelle said, placing her hand over her heart as a crowd of young women in New Hampshire watched in silence. "I can't stop thinking about this. It has shaken me to my core."

In the standard political playbook, Hillary's impulse made sense. The *Times'* Upshot predicted that her lead was insurmountable. She should sit back and let her opponent implode. Still, I wanted to see Beijing Hillary. I wanted to see Wellesley Bullhorn Hillary. I wanted to see the Hillary who defended her health-care plan to the House Committee on Ways and Means as "a mother, a wife, a daughter, a sister, a woman," and then compared Republican representative Dick Armey to Dr. Kevorkian. ("The reports on your charm are overstated and the reports on your wit are understated," Armey replied.) The Hillary I definitely did not want to see was the one who pivoted to a safe list of key Democratic constituencies.

"It's more than just the way he degrades women, as horrible as that is," she often said. "He has attacked immigrants, African Americans, Latinos, people with disabilities, POWs, Muslims, and our military." She might as well have added Diet Coke drinkers to the list.

Brooklyn kept complaining that the press cared more about the *Access Hollywood* video than the Russian meddling. They did this whole woe-is-me act about what "bad luck" they had that multiple intelligence agencies had released their report on Russian hacking on the same day as the *Access Hollywood* video. This spin had only one sensible response: *Give me a fucking break.* Clinton allies had tried for months to unearth damaging video or audio from Trump's reality TV years. In private, Robby didn't really care about the Russians. He thought Hillary should focus on the economy rather than try to explain to voters some convoluted intelligence report about a cyberattack.

There was no question that a foreign adversary attacking our democracy was the more significant news story—and that the media (myself included) should've given it more attention—but in terms of winning an election with a month to go, "Grab 'Em by the Pussy" was a slam dunk. The daily #PodestaEmail stories weren't helpful to Hillary (and I hated being summoned back to the newsroom to write them), but they mostly got buried in the vitamin pages as story after story on Pussygate and its implications for Trump landed on page one.

IN PRIVATE, TRUMP'S antics tore at Hillary. How could they not? On Thursday, she took a break from debate prep to have lunch in Manhattan with about a hundred donor friends at the St. Regis (cost: $33,400–$250,000). Meryl Streep introduced Hillary and wove a narrative yarn about standing up to bullies.

Trump had ended the first debate with a warning—"I was going to say something extremely rough to Hillary, to her family, and I said to myself, 'I can't do it. It's inappropriate. It's not nice.'" Hillary knew what he meant. The word "rape" had already come up in debate-prep sessions.

When Meryl sat down, Hillary took a sip of water and reminisced at length about another debate preparation session. The one when she was in the tenth grade at Maine South High School in Park Ridge. The debaters were boys, juniors and seniors, some of them bulky football players who had no time for the overachieving blonde. "But one football player stood out," Hillary said, her voice cracking.

He'd opened the old bifurcated window of the classroom, overlooking a courtyard where other students ate lunch and socialized. As Hillary tried to make her case, he broke off a rod that kept the windows hinged together and thrust it downward into the courtyard "like a spear," a show of strength and intimidation that stayed with her.

Fifty years had gone by but Hillary recalled the afternoon with such vivid detail that she started to sound as though she wasn't addressing donors at all but giving herself a stream-of-consciousness pep talk. Too bad only a hundred rich people got to hear it.

I DIDN'T GO to the second debate, or the one after that. The reporters who wrote the main stories typically watched from the newsroom. Before the second debate, we turned up the volume on CNN to watch Trump as he sat down for an impromptu press conference with three women, Paula Jones, Juanita Broaddrick, and Kathleen Willey, who accused Clinton of sexual harassment or sexual assault and one woman, Kathy Shelton, who at age twelve was raped by a man Hillary defended at a trial in Arkansas in 1975.

I had a hard time watching. The accusations in Roger Stone's book, deemed too salacious and unsubstantiated for us to write about a year earlier, now dominated the election. Stone always made the case that this wasn't just about Bill but about how Hillary and her liberal friends had treated his accusers, a sexual assault "Swift Boat" campaign designed to chip away at Hillary's biggest advantage—her standing among women voters. "She's not a victim. She was an enabler," Trump told Fox News.

During the primary, Stone's whisper campaign started to filter down to young women who grew up with the mantra that every woman who accuses a man of sexual assault deserved to be believed. Before the Iowa caucuses I'd written a piece about this generational divide among women. I heard that Lena Dunham, one of the campaign's top celebrity supporters, told a Park Avenue dinner party that she'd been disturbed by accusations that the Clintons and their allies dismissed and discredited these women. Hillary and the 1990s scandals, one young feminist activist told me, was "a big ball of ugly."

Hillary erupted after that piece, angry I'd written about the Lena anecdote (which her publicist said had been mischaracterized) and quoted other young women wrestling with this issue. Even my most sympathetic

Brooklyn sources told me that they understood why Hillary had been so "disappointed" in me, adding that they were "sad" about the state of the *Times*.

I didn't know what to think. It had been a different era in 1992 when Clinton campaign aides used words like "bimbo" and "floozy" and "stalker" to describe Bill's accusers. I once called Carville to confirm that he made the comment, "If you drag a hundred-dollar bill through a trailer park, you never know what you'll find . . ." His only complaint was that earlier reports mischaracterized the remark. "I was talkin' about Gennifer Flowers, not Paula Jones." Got it.

But blaming a wife for her husband's transgressions also seemed like the ultimate act of sexism. "Show me the wife who, when she finds out her husband is having an affair with a much younger woman, says, 'Oh, I feel such sisterhood with her,'" Katha Pollitt, the feminist poet and columnist for the *Nation*, told me.

And if these women had been violated by Bill and smeared by the Clintons and their liberal friends, weren't they being abused all over again now, trotted out like sexual-assault show ponies as 66.5 million Americans settled in to watch the second debate?

A reporter at the press conference yelled out, "Mr. Trump, did you touch women without their consent?" and Paula Jones shot back in her Arkansas drawl, "Why don't y'all ask Bill Clinton that? Why don't y'all go ask Bill Clinton that?"

We predicted Trump's move would backfire. That his bullying would lead to one of those times when the country saw Hillary's inner strength. Like in the 2000 Senate race, when Rick Lazio stalked toward Hillary's lectern, insisting she sign a pledge against soft money.

Or in the same debate, when the moderator, Tim Russert, asked Hillary how voters could trust her after she'd sworn in a TV interview that her husband had not had an "adulterous liaison" in the White House. Her eyes welled up, her lips tightened. Hillary replied, "I didn't mislead anyone. I didn't know the truth, and there's a great deal of pain associated with that."

Michael Barbaro and I spent all afternoon calling women voters. We declared that women watched the debate "through the same inescapable prism: a raunchy, three-minute recording in which Mr. Trump told of kissing and touching women however he pleased." We called this "Trump's new, agonizing and self-created reality" and declared his campaign "imperiled by his careless approach to gender . . ."

Less than a month later, Trump would win a majority of white women.

49

Bill's Last Stand

Bill Clinton was pissed off at the world. Trump parading out those women. Robby and his fucking data. The lawyers insisting he lay off dozens of employees and shutter the Clinton Global Initiative. Brooklyn blaming him for that smug James Comey taking over the email investigation.

Clinton turned seventy in August. He'd always thought Hillary would go to the White House and he could mostly stay in Chappaqua and continue to run the foundation. Even though Hillary moved back there after the State Department, Chappy had always been Clinton's stomping grounds. In the paid-speech boondoggle years, Hillary would sometimes spend nights at a suite in the Lowell on the Upper East Side. People said Chelsea needed a $10 million apartment so that her mom could sleep over, but that hardly ever happened.

"I hope I'll get permission to keep this foundation going," Clinton told Queen Latifah on her eponymous talk show in 2014. Wishful thinking.

Two years later and it was obvious there would be no way for Clinton to maintain any semblance of his postpresidential life. His ego took a hit. At Hillary's Roosevelt Island kickoff speech, I overheard a gaggle of young girls point at Bill as he made his way through the crowd and shout, "Look, there's Hillary's husband!"

The Clinton Foundation had been so dragged through the dirt that there was no way he could keep it going. I'd seen the foundation's work in Africa—seen deaf Ugandan children given the gift of hearing for the first time—and still helped fire the opening shot, with the 2013 investigation into mismanagement and dysfunction at the philanthropy (including the Yorkie) that I cowrote with Nick Confessore. The front-page story, my second on the beat, hadn't been bad for Hillary. It made Chelsea look gallant in her effort to professionalize her dad's charity in anticipation of her mom's arrival. But the story still provided Fox News with days of fodder and caused donors to panic. Clinton released an open letter pushing back against the *Times*.

Outsider Guy told me that by cowriting the story I'd single-handedly "taken AIDS medication away from thousands of kids." Not Nick. Not the *Times*. Not my editors. Not Doug Band, who'd built an enterprise off Bill Clinton Inc. But me. I dismissed this as his usual hyperbole, but three years later, after scrutiny of the charity and its top donors hadn't let up and its future looked uncertain, I realized he hadn't been all that far off.

Even if Clinton could somehow avoid the White House Easter Egg Roll and continue to run his foundation after the election, his philanthropy wouldn't have much money coming in. Along with shutting down the Clinton Global Initiative, the foundation said it would stop accepting foreign donations. There would be no more late nights with rich donors hanging on Clinton's every word as he yacked about soybean production in Rwanda. No more chartered international flights on which The Guys could charm donors' buxom wives.

Clinton had been so morose thanking the two hundred or so CGI staff members, many of whom would soon be out of a job, standing in the hub of the Midtown offices, slumped shoulders, bags heavy under his eyes, his gaunt frame swallowed under a navy-blue blazer, that Tina Flournoy, his chief of staff, prompted everyone to break out in "Happy Birthday." "I don't think I've ever done anything that I've loved as much as this foundation," Clinton said, the sad, lovable St. Bernard. "This is like a root canal for me."

But it wasn't just the foundation or the women. The entire election had become a repudiation of the Bill Clinton years—NAFTA, the 1994 crime bill, his dismantling of financial regulations, his gutting welfare by $55 billion. People—Democrats, even—linked the rise of Trump, the general degradation of the office of the presidency, to Clinton. He lied about getting a blow job in the Oval Office, so why not turn to a reality TV star who brags about sexual assault and seems proud to be propped up by the Russians?

It seemed the whole country was in open, angry revolt against a presidency that until recently most people would've said was pretty good. *What part of the 1990s didn't you like—the peace or the prosperity?* Incomes rose for everyone, not just the rich. No major wars, unless you count the Balkans and Somalia.

In the spring, Clinton had this embarrassing showdown with Black Lives Matter protesters. They waved signs like CLINTON CRIME BILL DESTROYED OUR COMMUNITIES and BLACK YOUTH ARE NOT SUPER PREDATORS, a term Hillary had used in 1996 to describe gang violence. Instead of ignoring them, Clinton extended a long finger and entered, as *Jezebel* called it, "Peak White Mansplain" mode.

"I don't know how you would characterize the gang leaders who got thirteen-year-old kids hopped on up crack and sent them out onto the street to murder other African American children," Clinton said, his bloodshot eyes bulging. "Maybe you thought they were good citizens. She didn't. She didn't." He raised his voice. He would've been at a yell if his voice had been stronger. "You are defending the people who killed the lives you say matter. Tell the truth."

Black leaders said it was a Sister Souljah moment for an entire generation. They weren't wrong.

As a proxy for his wife's primary campaign, which at that point was kept afloat almost entirely by black voters, the confrontation was a disaster. But the people who knew Clinton best said he was looking ahead to the fall, and especially the white working-class voters who put him in office in the first place. He figured that, for the most part, those voters would've agreed with him.

Not long after Hillary started her campaign, Clinton told the NAACP that his 1994 crime bill sent too many low-level criminals to prison "for way too long" and "made the problem worse." Even then, people saw only politics.

"I keep waiting for somebody to spring a trapdoor on me, oh look at Bill Clinton," I once heard Clinton say at a book party in New York that I wasn't really supposed to be at. "He really is still slick."

LIKE MOST REPORTERS, I loved covering Clinton, being around him, watching him work a room, even (or especially) at his most self-indulgent. He constantly disobeyed his staff and security detail. In White Plains, before the New York primary, he shook hands at a Hillary field office and then walked across a busy two-lane street. He blocked traffic and almost caused a multicar pileup with a single outstretched palm so he could drop in on the Dominican bodega across the street.

"Amy, Amy, take a look at this. This woman is Peruvian, but she has a Hindi tattoo on her hands," he said.

"Trump's America," I said.

Just then a tiny Mexican woman with a red purse sprinted across the street and dove into Clinton's arms. I don't know how she got past Secret Service except that she seemed hardly over four feet tall. Clinton hunched down and in a reflexive move, wrapped his whole body around her, both arms, his chin slumped down on her shoulder. Their cheeks touched. It could've been the last scene in *Dirty Dancing*, except with an elderly Mexican woman and a white-haired former president.

By the fall, Clinton was tired of reaming Robby out on conference calls, screaming that the campaign shouldn't ignore the white voters he'd won in '92 and '96 and whom Hillary won over in upstate New York in the Senate years and in her 2008 campaign.

Clinton tried to tell them that he knew Trump better than any of the other candidates did. Hillary, thinking Trump was a bigger donor than he actually was, had insisted they attend his 2005 wedding to Melania Knauss, despite a couple of aides warning her not to go. Hillary ended up sitting behind Shaquille O'Neal at the ceremony and could hardly see anything except the ninety meters of white satin tulle of Melania's Dior gown pass down the aisle.

Clinton thought Trump's antitrade economic message could be lethal.

He said something like "This guy is making Hillary look like the elit-ist" almost daily. Nobody reminded Clinton that his sexual past left the campaign paralyzed in hitting Trump's treatment of women. Or that shouts of "Bill Clinton's a RAPIST!" now interrupted almost all Hillary's rallies.

The campaign treated Clinton like a distraction, a gifted but prob-lematic child who needed to be kept busy. Regular updates to Brooklyn from Clinton's team included, "Fair to say we didn't break anything." By October, Clinton had splintered off from the campaign, venturing on a series of "Stronger Together" bus tours. He went places Hillary wouldn't go to—tried to talk to the Bubba-Trump continuum. North Florida, eastern Ohio, the Mahoning Valley, Iowa, rural Pennsylvania, stops in Wisconsin and Michigan. At a diner in Buffalo, Clinton ate a french fry off a customer's plate (with permission).

In East Africa, I wrote in my notebook the Zulu greeting Clinton used wherever he went, *sawubona*, which means "I see you," to which villagers would reply *ngikhona* or "I am here." He felt the same way about Trump voters. "An enormous number of people feel like they're not seen, that nobody gets them," he said.

"You can actually go places where you can make a difference in the vote because people don't expect you to show up there," Clinton said. "I found it was the most efficient use of the things I could do for the cam-paign, but I also just like it. It's much more like the ways I campaigned when I was a young person starting out. When I was out at nineteen years old working for other people."

But if there was one thing Clinton would remind you, and keep remind-ing you until you got it, it was that he wasn't nineteen anymore. This came out in all types of weird ways. At a black church in Detroit, he'd had body envy. "Anytime I watch an NBA basketball game, I think, if I had a body like that I would've gone into a different line of work," Clinton said.

On a stop at a Mexican restaurant in Santa Fe, a twenty-four-year-old Bernie supporter, Josh Brody, ripped into Clinton about the welfare overhaul and Wall Street.

"It's a nice little narrative," Clinton said, hovering over his booth at Tia Sophia's.

"The Democratic Party is now a party that's supposed to represent the working class without many working-class constituents," Brody said, blaming the shift on Clinton moving the party toward big business.

"If you never have to make a decision, then you can go back to the past and cherry-pick everything [for a] narrative that is blatantly false. What you're saying is false," Clinton said.

He could've walked away. His aides kept urging him to walk away. ("Other people are waiting," his body man, Jon Davidson, said. "I think we're gonna agree to disagree here, guys.") But Clinton wasn't finished.

"People like you show up in presidential elections. A lot of them stay home. That gives you the benefit of being able to criticize everybody," Clinton said. His aides were physically nudging him to the next booth. "If the best thing to do is just say no and lob bombs, you don't get anything done."

I didn't believe the rumors that Bill was sick or losing it. My Irish power broker friends—the same ones who invited me to a luncheon the year before at which Hillary sat next to Gerry Adams ("palling around with terrorists," a source texted me)—invited me to a St. Patrick's Day event with Clinton. Mouths fell open as he extemporaneously wove a speech decrying political polarization into a crescendo with flavors of Yeats.

"We can never let our hearts turn to stone, and we can never let things fall apart so much that we cannot build a dynamic center where the future of our children counts more than the scars of our past," he said.

But something had changed in Clinton. He meandered through his usual remarks, telling his extremely elderly, almost entirely white crowds how Hillary was "the best damn change maker." And he spoke with a kind of schmaltzy desperation I hadn't heard before. "You guys did well when I was president, let's come in and talk," he told a couple hundred people at a rally in Fort Collins, Colorado. He raised his dry voice to speak over the hum of a turboprop plane circling overhead with the words GO TRUMP painted on its wings.

In Pueblo, Clinton told the crowd, "Look, his base is where I grew up. I was born in Arkansas to a mother of Scots-Irish lineage. These are good, honest people," he said, "but always and forever we've been manipulated by scoundrels." In the back, a man shouted, "Lock that bitch up!"

50

Chekhov's Gun

NEW YORK CITY, OCTOBER 28, 2016

File under headlines that didn't hold up: STOP TRYING TO MAKE ANTHONY WEINER'S SEXTING A POLITICAL ISSUE, *New York* magazine, August 30, 2016.

The day October delivered its final big surprise, my colleague Mike Schmidt was visiting from DC. He sat in the cubicle next to me in the newsroom as we both worked our sources. Twenty minutes after the Clinton campaign announced in a show of confidence that Hillary would hold an early voting rally in Arizona, a state that had gone red in eleven of the past twelve presidential campaigns, but seemed potentially in play, news broke that James Comey sent a letter to Congress stating the FBI had found additional emails related to Hillary's private server. Trump wasted little time declaring, "This changes everything."

Schmidt heard the emails had been unearthed during a separate investigation into Anthony Weiner's sexting with an underage girl. He kept yelling into the phone, "They've got Weiner by the balls!" until I finally G-chatted him that he had to stop saying that.

The *Times* news alert went out that the emails had been found on a computer Huma had used. The Weiner connection both was unbelievable and yet, in some sad way, made perfect sense: Hillary, married to an alleged sexual predator, could lose to Trump, an alleged sexual predator, because of Weiner, an alleged sexual predator.

No one understood why Huma had stayed with Anthony for so long, but the best explanation I'd heard came from a close girlfriend of Hillary's who reminded me that Huma had grown up amid the tumult of the Clintons' own marriage.

"You can't blame her," this friend said, using the same excuse The Guys had used for Chelsea. "She was raised by wolves."

I thought back to 2013 when I first heard about the "Carlos Danger" scandal, to the stories I wrote about The Guys hoping to contain Huma's personal life so that it didn't spill into Hillary's political future. They protected Huma as if she were a beloved little sister and a vital append-age of Hillary. Big donors were less sympathetic, imploring Hillary to put Huma in a less visible role. At least one top donor confronted Huma directly, in 2013, pleading with her, for Hillary's sake, to step down. "I'm good at what I do and that's Hillary's decision," Huma replied.

Now, in the last act, with eleven days before the election, Huma's problems exploded in one final self-inflicted, seismic wound.

"It's like Chekhov's gun," I said as we stood around discussing the news.

A colleague who overheard this said, "I didn't know they knew who Chekhov was in Texas."

Very Senior Editor came by my desk to ask, "She's not gonna lose, right?"

I gave my extremely professional assessment of the situation.

"Brooklyn is freaking the fuck out," I said. "Her trust numbers are already shit."

In August, after the POP GOES THE WEINER cover in the *New York Post*, Trump told us, "I only worry for the country in that Hillary Clin-ton was careless and negligent in allowing Weiner to have such proxim-ity to highly classified information. Who knows what he learned and who he told? It's just another example of Hillary Clinton's bad judgment. It is possible that our country and its security have been greatly compro-mised by this."

His statement had seemed so far-fetched that Pat Healy and I took a fair amount of outrage from the #ImWithHer contingent for including it in a front-page story ("THIS CHANGES EVERYTHING": DONALD TRUMP EXULTS AS HILLARY CLINTON'S TEAM SCRAMBLES). But Trump had been half-right. The FBI didn't find any additional classified or incriminating emails on Weiner's computer, but the "bad judgment" line stuck.

Hillary was en route to Cedar Rapids when the news broke, accompanied by her childhood friend, Betsy Ebeling, a sweet gray-haired Midwesterner whom the campaign rolled out every time they needed a testament to Hillary's warmth and down-to-earthiness, and the celebrity photographer Annie Leibovitz. Robby Mook had been on board to brief the Travelers about Hillary's trip to Arizona and how she'd expand the map. Hillary didn't initially see the news—nor did most of the press—thanks to the plane's shoddy Wi-Fi.

When the Stronger Together Express touched down, disbelief, followed by alarm, spread throughout the front cabin. The Travelers bustled onto the tarmac hoping to scream a question, "SECRETARY! WHAT ABOUT THE FBI?" Hillary lingered on board. She had the photo shoot with Annie Leibovitz to finish. She'd later tell friends that the development was "just another crisis" in a career full of them.

In the newsroom, we turned up the volume to watch Hillary's brief press conference that evening. Part of me longed to be there shouting questions myself.

But mostly, I thought of Sara.

I'd spent the past year bringing chocolate babka and challah loaves to Sara Ehrman, the feminist firebrand whom Hillary had lived with after law school when she worked on the Watergate Committee. Forty-two years earlier, in August 1974, Sara drove Hillary, then twenty-six, to Fayetteville, Arkansas, to be with Bill Clinton. Sara tried to talk her out of the move the whole way down. "We'd drive along, and I'd say, 'Hillary, for God's sake, he'll just be a country lawyer down there.'" And each time, Hillary would answer the same way, telling Sara, "I love him and I want to be with him."

Sara was ninety-seven but feisty, still dispensing tough love to her most famous protégé, Hillary, and a revolving door of Washington women who came to her sunny Kalorama apartment, bearing gifts and seeking career advice. We'd become close over the many afternoons I'd try to woo her into talking on the record about the two-day 1,193-mile journey that changed Hillary's life. For over a year, Hillary had turned down my many interview requests to do a piece on their relationship, and Sara remained reluctant. After the election, Sara showed me emails from Brown Loafers instructing her not to talk to me, basically saying that I hated Hillary and couldn't be trusted to be fair—a warning Hillary had asked him to pass on. But Sara finally agreed to talk to me anyway, writing back to Brown Loafers something like "For God's sake, she's just a nice Jewish girl from Texas."

Sara reminisced about Hillary's sloppy room, with brown clothes and books and a bicycle strewn about. She told me about the deeply personal conversations they shared on the two-day drive down Interstate 81 in Sara's beat-up '68 Buick sedan ("an old rattletrap," she called it).

The journey, I wrote in the *Times*, had "the ingredients of a classic American road trip—a cheap motel, tchotchke purchases, encounters with drunken strangers and deeply personal conversations" and offered "a glimpse of Hillary the public seldom sees . . . wide-eyed and eager, vulnerable and afraid, at the cusp of a momentous decision that would alter the course of her life."

The road trip story—and accompanying video interview with Sara, sitting on the sofa in a sea-foam sweater set that brought out her eyes—was my favorite article that I ever wrote on the beat, maybe in my entire career. It was published on the *Times* website hours before news of the Comey letter broke. Hardly anyone read it. The story had been scheduled to run prominently on the next day's front page, but never even made it into print. Several months after the election, I would write Sara's obituary. Hillary told the story of their road trip at the memorial service.

The Comey news would lead the entire front page—three stories, seven bylines (including mine), a four-column photo of Hillary, Huma standing over her shoulder arms akimbo. The layout would live in infamy, proof to Hillary and the #StillWithHer crowd that the *Times* blew the email story out of proportion, the climax of its decades-long anti-Clinton vendetta.

"I am confident whatever they are will not change the conclusion reached in July," Hillary told reporters when barraged with questions that afternoon about the newly unearthed emails. Asked about the Anthony Weiner connection, Hillary said, "We've heard these rumors, we don't know what to believe."

Less than four minutes later, she closed her green binder and turned to leave.

The Travelers yelled, "Secretary, is this going to make your campaign so much harder?"

"Secretary, how did you learn about this?"

Hillary pointed at the scrum. "Same way you did, from the press."

"Are you worried this could sink your campaign, Secretary?"

Hillary, now almost out of sight, shook her head and laughed.

51

Hillary's Death March
to Victory

You are in Akron, Ohio, for an Ohio Democratic Party
Voter Registration event with Secretary Hillary Clinton at
the Goodyear Hall and Theater on Monday, October 3,
2016.

Hillary reminded the audience that LeBron James had endorsed her.
"Now, I may become president, but he will be king of Ohio for as long
as there is a king," she said.

Dozens of empty chairs sat in the press area. Extension cords dangled
unused off folding tables. Cherry pickers set up to give photographers
an aerial shot of Hillary sat idle. I had to remind myself that there was a
time, during the Harkin Steak Fry, before Hillary was a candidate and
while Trump was still a reality TV star, when *she* had been the media's
obsession. Two hundred reporters had stampeded across the lawn for
a glimpse of the most irresistible, dramatic story of the "horribly dull
political year to come." Hillary, the would-be candidate able to capture
the world's attention with a single flip of sirloin, now hardly registered.

Pretty soon we were calling it Hillary's Death March to Victory. I
don't know how else to explain it except to say that it didn't feel like a
winning campaign. Hillary went through the motions.

"Hello [insert swing-state city here]!"

Eight years earlier, I'd experienced an actual winning campaign, and not just some winning reelection campaign, but a spectacular, holy-shit-America-is-going-to-elect-a-black-man winning campaign. I saw the euphoria—the millions who never thought elections mattered to them, standing in line for hours just to get a glimpse. You never get used to seeing a crowd of a million people. Every time we walked into a rally, in stadiums mostly, my jaw would drop. Obama must've felt the same because he never wanted to get off the damn stage. He'd look out at the crowds, the HOPE signs, the tears in people's eyes. He'd ramble on for fifteen or twenty minutes, interrupted with frequent shouts of "I love you," and his cool, "I love you back." He knew that was the best it would ever get.

Hillary's campaign did not feel like that. In fact, if there was a single unifying force behind her candidacy, it was her obvious desire to get the whole thing over with. "This election is ten days away," she said at a rally in Des Moines. "Eleven, but we're more than halfway through today."

Aside from the local touches, each speech, crowd, city blurred into one. The women, mostly boomers, who wore NASTY WOMEN FOR HILLARY buttons. The senior citizens who sat in folding chairs to one side of the stage, their walkers or canes resting nearby. The gay men who had blue *H*'s painted on their cheeks. The LOVE TRUMPS HATE signs—even her catchiest slogan was about Trump. Katy Perry's "Roar" and Rachel Platten's "Fight Song" filling the high school gyms with the scoreboards set to TIME: 2016, HOME: 45, AWAY: 45. The "Lock Her Up!" protesters outside. The scent of the nearby porta-potties.

The Travelers didn't know if we were in Akron or Toledo, Cleveland or Cincinnati, Raleigh or Charlotte, so the press advance kids started to tape helpful reminders to our workspaces in each city.

You are in Toledo, Ohio, for a Secretary Hillary Clinton Economic Speech at Dr. Martin Luther King Jr. Plaza on Monday, October 3, 2016.

"Did you see the last debate?!" had replaced "Deal me IN!" as Hillary's favorite line.

She was so far ahead in the polls after the third debate that I wrote lines like "Mrs. Clinton is likely to prevail against Mr. Trump in two weeks," and "Mr. Trump's party increasingly concedes he is unlikely to recover in the polls."

Brooklyn was so cocksure that they turned to Senate and congressional races. The campaign, flush with $153 million in cash, spent $1 million on voter turnout in Indiana and Missouri. Robby vowed to "dramatically" expand efforts in Arizona, spending an additional $2 million on advertising and sending their most valuable (and reluctant) campaigner, Michelle Obama, there. The campaign bought ads in Texas and Utah. Organizers in Michigan and Wisconsin still pleaded for resources, but Robby saw opportunities to "expand the map" and "make sure Democrats controlled the Senate."

This new mission made Hillary's stump speech even more stilted. In Lake Worth, Florida, the crowd sang "Happy Birthday" to Hillary, who turned sixty-nine that day. "Thank you so much for singing to me," she said. "I hope that one of the best gifts that you can give yourselves would be sending Patrick Murphy to the United States Senate."

Adding to the gloom of superiority in the campaign was the angst about WikiLeaks' daily release of more #PodestaEmails. Each dump brought fresh conspiracies, including right-wing theories that a reference to "Madre" in one email ("please don't burn the source or Madre may pay the price") was a code name for a secret CNN source trying to help Hillary. In reality, Madre is one of The Guys' pit bull rescues.

With weeks left, I was dying to spend as much time as possible on the road, but the email dumps often tethered me to my desk. I'd been on my way to the White Plains airport to fly to North Carolina, where Hillary and Michelle Obama would have their first and only joint rally, only to be summoned back to the newsroom by my editors to read through a new pile of emails related to the Clinton Foundation.

This election isn't happening in my cubicle.

I traveled on weekends, holidays. I went to a black church in Charlotte with Hillary on Rosh Hashanah. I'd had to scramble to find the *Times* editors on "goy duty."

> You are in Cincinnati, Ohio, for a rally with Secretary
> Clinton at Smale Riverfront Park on Monday, October
> 31, 2016. Note: Happy Halloween!

Early one Sunday morning, the Travelers, who'd spent the night at a Marriott in a research park in Durham, North Carolina, rode over to Hillary's hotel, a luxury spa with Zen gardens and outdoor fire pits overlooking a thick forest of Carolina hemlocks and magenta rhododendrons.

We barreled into the earth-hued lobby spritzing ourselves with raspberry-scented lotion in the gift shop, digging underneath the suede chaise longue chairs for power outlets and generally destroying any shred of Zen that the hotel had once offered.

"The Everydays don't stay here."

"No, no they do not."

"Look! Is that free coffee?"

From there we went to a black church in Durham (*This is the day the Lord hath made . . .*) followed by a couple early-voting rallies. We got back to Westchester after 10:00 p.m. I hadn't written a single word, but I did take a stab at TLC's "Waterfalls" on a karaoke machine a photographer had set up on the plane.

Thanks to Hillary's celebrity endorsements, the Travelers spent almost every night at a free concert—Katy Perry (Philly), Adele (Miami), Elton John (New York), J.Lo and Marc Anthony (Miami), Chance the Rapper, Jay-Z and Beyoncé plus backup dancers in blue pantsuits (Cleveland).

The wire reporters sat staring into the glowing eyes of their laptops during the Beyoncé–Jay-Z concert. In khaki pants and sweater sets, these DC reporters looked about as incongruous in the jubilant almost entirely black audience as Mike Schmidt waddling onto Waikiki Beach. Annie and I stood up and danced. I was anticipating the unflattering photo and the Breitbart headline FAWNING PRESS GOES WILD FOR HILLARY IN CLEVELAND, but I didn't care.

And when Beyoncé displayed, in a multimedia explosion of hip-hop and girl power, the phrase that twenty-four years earlier had cast Hillary as Lady Macbeth and an affront to stay-at-home moms, I even choked up. "I suppose I could have stayed home and baked cookies and had teas, but what I decided to do was to fulfill my profession," flashed on the screen. Signed, "Hillary Clinton."

It was a major milestone. Maybe Hillary really would win. Of course, when I asked Jen if that meant Hillary had reclaimed the cookies comment, for which she spent twenty years apologizing, Jen said Beyoncé had creative control and Hillary hadn't known about it. "Yeah, that was, uh, interesting," Jen said.

I looked around at the thousands of pumped young voters who poured out of the Cleveland arena. They didn't look anything like Hillary's usual crowds. For a second, I thought they'd deliver Ohio. But then I thought back to Iowa and the Demi Lovato concert and what Pete D'Alessandro had said: *They just came for the ribs.*

In the final weeks, I went on the road even though I knew it meant I'd miss out on bylines. In meeting after meeting, on daily 8:00 a.m. conference calls, colleagues who covered Trump explained that there was no chance in hell he could win. I agreed, thinking this was just how Hillary would win, the long-suffering feminist heroine who would make history not in a festooned lovefest but in a dreary, mechanical slog.

Robby had just done a conference call with donors, telling them Hillary was up by seven points in Michigan, twenty points in Wisconsin, with 10 percent of the early vote tallied. Hispanic turnout was up 139 percent in Florida. She'd win North Carolina by three or four points. "Comey did not change the fundamentals of the race. We saw tightening before Comey, and there's no dramatic shift in our polling," he said. "It's the tightening of the GOP coming home."

Hillary joined the call. "We know Arizona is a stretch," she said, noting that she had sixteen thousand people at her rally in Phoenix. "We're doing all we can to turn it blue."

It wasn't mathematically possible that Trump could win, and yet I still couldn't kick the feeling that Hillary wouldn't win either.

You are in Pittsburgh, Pennsylvania, for a Secretary
Hillary Clinton Democratic Organizing Event at the
Great Hall at Heinz Field. It is Friday, November 4, 2016.

In Pittsburgh, the Travelers had the run of Heinz Field while Hillary and Mark Cuban (she loved having her own bombastic billionaire turned reality TV star along) held a rally in an indoor pavilion lined with Steelers memorabilia. It was wonderful to be outside, not stuck on a bus or in a high school gym on a sunny, sixty-five-degree fall day. We raced out of breath to the fifty-yard line. It was just ten of us, the ones who'd covered Hillary the longest. We looked out at the sun refracting off sixty-eight thousand empty orange seats. "Classic HRC rally crowd," Annie said. "Just kidding. Just kidding."

The front sanctum of the Stronger Together plane started to look like Hillary's West Wing—longtime friends, the loyalest of the loyalists, rejoined the entourage, including Cheryl Mills, Maggie Williams, Capricia Marshall, Huma (back from a brief banishment in Brooklyn), and OG, who'd been mostly kept in his padded room (or DC office) during the campaign. "Capricia Marshall, will you please report to the principal's office?" one of The Guys said into the intercom, summoning her to Hillary's chambers.

For all the talk early on of bringing in fresh talent to run her campaign team, Hillary would end it with the same box of broken toys who'd enabled all her worst instincts since the '90s. I emailed Carolyn, "I'm waiting for Sid Blumenthal to join us."

IN 2015, BEFORE she started her campaign, Hillary talked to aides about what she called her "Al Gore problem." If she was going to be herself, substantive, prepared, prone to policy talk, she knew she'd also come off as stiff, a square. In these final weeks, she finally seemed at peace with that.

She did a whole riff on making lists. "I have a plan for just about everything . . . You know, maybe this is a woman thing. We make lists, right? I love making lists. And then I love crossing things off!" She'd build "Get Out the Vote" rallies around her college calculator. ("You can actually go to hillaryclinton.com/calculator to see how much money you and your family could save with our plan.") Speaking to college students, she'd give them "a little homework assignment." ("If you add up the number of jobs that our economy added when Barack Obama and Bill Clinton were president and if you compare it with the two Bush presidencies and the Reagan presidency, you'll see what I mean.")

In Grand Rapids, Hillary led a call-and-response about interest rates. "There are so many people paying eight, ten, twelve, higher. Now where did that—how much?" She cupped her ear to hear a few numeric shouts from the crowd. "Twelve-point-seven percent! Fourteen-point-one percent? I gotta tell you, that is outrageous."

She kept making references to *Hamilton*, even though tickets to the Broadway hit would've been prohibitively expensive for most Everydays. "Wow, I'll tell ya, I've seen it three times and I listen to the score all the time," Hillary told a crowd at an outdoor rally in Charlotte.

I lifted the right side of Peter Nicholas's noise-canceling earmuffs and said, "It didn't end well for Hamilton."

But more than any data or talk about Democrats retaking the Senate, I knew Hillary thought she'd be president when she brought back the 3-D printer. "I am not one of those folks who think, 'Well, we just can't make it in America anymore' . . . We can lead the world in precision machining, in 3-D printing!"

She almost only did call-in interviews with black and Hispanic radio shows, and she even resisted doing those. She seemed uninterested, botching the softest of softballs. Hillary told DJ Envy of *The Breakfast Club* that she loved to dance, "and so any chance I get, I will dance. I'm not sure that it would be anything that you'd be saying was good dancing but—"

"You don't do the robot and stuff like that?" DJ Envy asked.

"No, I'm not a robot—I don't do that robot stuff, yeah."

When Sam Sylk, another black radio jockey, welcomed Hillary and said, "Yeah, I'm pumped up on why to vote," Hillary laughed and said, "Well, that makes one of us."

"Yeah, makes one of us, huh? Okay."

She did so many radio call-ins that I asked Brown Loafers if Hillary would call into Michael Barbaro's new podcast, *The Run-Up*. "The *New York Times* podcast audience isn't our target demographic," he replied.

USUALLY BY THE final stretch, candidate reporters are so brainwashed from living in the bubble that we all believe our horse will win even if the facts say otherwise. Think about it: For months, years sometimes, we've only talked to die-hard supporters. We've only heard spin from one side. We've been at hundreds of rallies where the only voters we see want our assigned candidate to win. Traveling in the motorcade means the campaign rushes us past the occasional protesters, a passing glare that has faded by the time we're captive on the bus again.

Romney's traveling press fell for it hard in 2012; so did the McCain-Palin reporters in 2008. In the waning weeks of the 2008 primary fight against Obama, when Hillary added more events to her already packed schedule and party leaders urged her to drop out, I still believed she had a chance. *You just don't understand; the polls don't see what I see.*

In 2016, the opposite happened. The Hillary press and the Trump press both thought we were the sad sacks doomed to cover the losing candidate.

52

The Tick-Tock Number One

NOVEMBER 7, 2016
11:10 A.M.
    Traveling press and pool wheels down Pittsburgh
    International Airport Atlantic Aviation FBO.

I had my ideal aisle seat for the final leg—in the second row of the press cabin next to Annie in the middle seat and Ruby at the window. Varun (or Arun, as we all called him after Hillary's flub), who assigned us our seats, had come through. Either that or he just didn't want to deal with my diva fit if I hadn't gotten the aisle. We loaded off the plane at our first stop, suddenly aware of where we'd landed and how strange it was, given that Hillary was so far ahead in Pennsylvania.

"Does anyone know why we're back in Pittsburgh?" a Traveler asked.

"Yeah, weren't we, like, just here like two days ago?"

". . . and in Philly yesterday?"

"Yeah, why couldn't we have like stayed in Philly and gone straight to Pittsburgh?"

On Sunday, we'd gone to the Mt. Airy Church of God in Christ in Philly where backstage Hillary and the gospel singer BeBe Winans belted out "Amazing Grace" so loudly that her voice echoed through the closed doors.

I remembered what Robby said the other day, about how the campaign's analytics showed Hillary would do better in the Philly suburbs than any Democrat had in decades.

"Turnout, obvs," I told the Travelers. I work for the *New York Times*, I'm Number One.

"Yeah," they agreed.

"She wants to kill him in Pennsylvania."

1:05 P.M.

You are in Pittsburgh, Pennsylvania, for a Secretary Hillary Clinton Get Out the Vote event at the University of Pittsburgh on Monday, November 7, 2016.

For months, Hillary had been telling voters that what the country really needed—in addition to a $275 billion infrastructure plan and a revised corporate tax code—was more "love and kindness." But there'd been little of either this election year. She'd spent the past several months pillorying Trump as "unfit" and "dangerous," a "loose cannon" and a "puppet." I thought back to the day before the 2008 election when I was the pool reporter in Obama's motorcade riding over to Grant Park after the results came in, and how optimistic the country felt then. David Plouffe, Obama's 2008 campaign manager, summed up 2016 as, "Hope and change, not so much. More like hate and castrate."

Now, with hours left before the polls opened, Hillary stopped making it all about Trump. "I'm here to ask you to vote for yourselves, vote for your families, vote for your futures, vote on the issues that matter to you because they are on the ballot, not just my name and my opponent's name."

"I love you!" a group of young girls shouted in unison.

It wasn't that, in Hillary's 575 days on the campaign trail, adoring fans hadn't professed their love for her. They had. But in Pittsburgh, Hillary looked as though she finally believed them. She paused, tilted her head to the right, and looked at the girls. Abandoning the teleprompter, she said, "I love you all, too."

5:05 P.M.

You are in Grand Rapids, Michigan, for a Secretary Hillary Clinton Get Out the Vote rally at Grand Valley

State University Fieldhouse. It is Monday, November 7,
2016.

"LOADING!" a press aide yelled. Hillary had just wrapped up her second and final rally in Michigan since she lost to Bernie back in March.
Two rallies down, three more stops to go.

I walked through the parking lot, my laptop open in my arms even
though I wouldn't file my story until the end of the day's swing, around
3:00 a.m. A lone protester held a sign LIAR, LIAR, PANTSUIT ON FIRE.

I turned my head to the right and saw Hillary's staff van. The sliding
door was open and inside, spread out longways, his legs on the upholstered bench, shoes practically jutting into the sun, was OG, the Original Guy, who years earlier had suckered me with access and charisma
and then mindfucked me into believing my colleagues would destroy
me and stomp on my cold, irrelevant corpse before we even got to Iowa.
The one whose email manifestos had ruined so many weekends. The
one whose good side I should've tried to stay on, but stupidly chose to
provoke, Polar Bear–style. The one who had no idea that the next day
he'd watch his dream job in the White House slip away to Hope Hicks,
a twenty-eight-year-old former Ralph Lauren model.

I'd wanted for so long to prove to him, and by extension to Hillary,
that I could survive the Steel Cage Match. We locked eyes. I smiled and
kept walking.

8:02 P.M.
    Traveling press and pool depart Philadelphia
    International Airport en route Independence Hall.

I held my phone up to the speakers trying to capture Bruce Springsteen's
raspy lyrics as they filled the air outside Independence Hall. In college,
Bobby spent summers on a J-1 visa working as a carny on the Jersey
shore. He had bleached-blond hair and operated a balloon ride on the
boardwalk in Wildwood. He tells the story of his friend's mum seeing

this horde of gangly Irish schoolboys off at the Dublin Airport to America. "Mind the AIDS!" she yelled at them as they boarded.

I'm more of a Tom Petty person, but for Bobby nothing epitomizes America and why he wanted to move here and become a citizen more than Springsteen. He was so excited to vote for the first time on Tuesday. I wished he'd been there to hear "Long Walk Home" performed live, outside, to thirty thousand people hours before the election. Springsteen called the song "a prayer for postelection," and declared that on Tuesday, Trump's "ideas and that campaign are going down."

I called Bobby several times, but it kept going to voice mail. I recorded a few minutes of music, and then said, "I love you so much. Thank you for everything. I wouldn't be here tonight without you. You're everything," and hung up.

9:11 P.M.
> GOTV rally on Independence Mall with Hillary Clinton, President Obama, Michelle Obama, President Clinton, Chelsea Clinton, and Jon Bon Jovi. PLEASE NOTE: THIS EVENT IS OUTSIDE.

Hillary and Obama came onstage arm in arm. The sky had cleared and a sliver of a half-moon glowed over Independence Hall. They waved and pointed at the crowd, Hillary's biggest yet. These two Democrats whose rivalry had been the backdrop of my reentry into American politics now stood as one unit—Stronger Together, or something like that.

12:52 A.M. (Election Day)
> Midnight Get Out the Vote Rally, Reynolds Coliseum, North Carolina State University, Raleigh.

> Dear HFA Traveling Press:
> Congratulations on making it to the LAST RALLY of this

campaign! We hope this Midnight Get Out the Vote Rally is
a memorable last stop. We've enjoyed working with you lots!
　　—Press Advance

You can count on Hillary's "midnight" rally starting closer to 1:00 a.m.
Bon Jovi decided to tag along on the flight to Raleigh-Durham. We saw
him boarding, a silhouette in black who waved to the press. When we
landed and walked into the Reynolds Coliseum at North Carolina State
University, Lady Gaga, in black aviator glasses and a high-collared velvet
brigandine with a red armband, sat at a grand piano singing, "I want
your love."

You could tell the campaign had screwed up because the narrow pas-
sageway they'd carved out for the press, in between the throngs of college students, put us so close to Gaga we could've grabbed her as she
stood on the piano for the finale of "Born This Way."

"This way, traveling press! Keep moving!" a press aide yelled as we
stood and gawked at the pop star.

Bon Jovi and Gaga did an impromptu duet of "Livin' on a Prayer."
Gaga slid off her glasses and told the crowd, "I could never have fathomed
that I would experience in my lifetime that a woman would become president of the United States," and the six thousand people who seemed un-
fazed by Hillary's tardiness exploded. Everywhere I looked, I saw women
of all ages, most of them of the Nasty variety, crying as they watched
Hillary onstage.

She joined hands with Bon Jovi and Gaga. This odd threesome thrust
their arms overhead. Hillary couldn't help herself. "Until the polls close
tomorrow, we're gonna be livin' on a prayer!" she said. The Travelers
rolled our eyes at each other and giggled. Would she still make pop-
culture puns as president?

The campaign must've relaxed its corporate-issued poster rules because everywhere people held handmade signs that said IT'S TIME! and
HILLARY 4 ME. A chant of "I believe that she will win! I believe that
she will win!" broke out. At that moment, I believed that she would
win, too.

1:55 A.M.

> Traveling press and pool wheels up Raleigh-Durham
> International Airport en route Westchester County
> Airport.

The next hour had the feel of an in-flight victory party. The flight attendants could hardly get us all to sit for takeoff, and then everyone, including Bill, Hillary, Chelsea, and a cabin packed with their closest aides and
friends, bounced out of their seat belts and poured into the aisle.

Bon Jovi ventured to the back cabin where a couple of Travelers asked
for selfies. I scooted around the fangirls who encircled him and saw OG.
We talked as if we had no history. As if nothing had passed between us.
We were two acquaintances who ran into each other in the grocery store
checkout line.

"How have you been?" I asked.

"Not bad, and you?"

I don't remember how the Russians and WikiLeaks came up, but
when it did, he said something like, "Can you imagine the epic scandal
if the *Times'* emails were hacked? It would make Jayson Blair look like
nothing."

I nodded and agreed, but I wasn't sure why he was telling me this.
Then he dropped that he knew the cybersecurity firm the *Times* hired
to secure our servers. "Nice guys over there. I'm friends with a couple of
them . . ."

"Well, okay, nice seeing you," I said, and I walked back to my seat, a
pit in my stomach knowing he'd soon be running the country.

3:45 A.M.

> Traveling press and pool wheels down Westchester
> County Airport, Ross Aviation.

Because Hillary didn't want the day to end and wanted to torture the
press with one final rendition of "Fight Song," Brooklyn assembled a

couple hundred of her most loyal neighbors, staff, and supporters for one last soiree on the tarmac in White Plains. Hillary stepped off the Stronger Together Express for the last time, wearing her thick glasses. I stood on the platform set up for the photographers and watched her disembark. I thought about her 70 percent approval rating when she was secretary of state, the last time she'd regularly worn glasses and scrunchies and a multicolored coat she bought at a market in Afghanistan in the '90s. The country liked Hillary then. Maybe they could like her again when she became the FWP. She spent five minutes shaking hands and stepped into her van for Chappaqua.

53

The Tick-Tock Number Two

Hillary's motorcade arrived at the Peninsula hotel, a limestone and steel tower on Fifth Avenue with a mint-colored cornice as resplendent and dignified as a jade Buddha. Only family and her innermost circle were invited to watch returns come in at the Peninsula, before everyone headed across town to the Javits Center. Hillary stepped out of her van believing that when she got back in later that night—probably around 10:00 p.m., maybe even earlier, her aides assured her—she'd be the president-elect. She strode into the Peninsula's amber-lit lobby, with its scent of fresh hydrangea and Chanel No. 5 and a row of attendants in white gloves who looked ready to pantomime. Just around the corner, steel barricades set up along Fifth Avenue fenced in a small crowd of Trump supporters, probably from Staten Island, who chanted "Drain the swamp!"; a couple of protesters who held Andy Warhol–like silhouettes of Trump with the words PRO RAPE; and enough NYPD officers in riot gear to stop the Tiananmen Square massacre.

If I had to choose a single symbol of Hillary's outsize optimism that night, it wouldn't be the two-minute fireworks display over the Hudson River that would've burst over the Javits Center's glass-domed ceiling (until the Coast Guard put the kibosh on the idea), or the speech devoting her victory to Dorothy, which she had recited over and over to make sure she didn't cry as she had in her Brooklyn speech. It was her choice of hotels.

For as long as I'd covered the Clintons, they'd been savvy enough to know that New York Democrats hold major events at the sturdy Times Square Sheraton, whose workers are proud AFL-CIO members. Whether you're rich or rank-and-file, if you're a Democrat and you're in Manhattan, chances are you've flooded the hotel's hallways and narrow elevator banks for one cause or another. Al Sharpton held his annual convention of civil rights leaders there, and for years the Clinton Global Initiative had transformed the Sheraton's gloomy ballrooms into what felt like a multiday walk-through Clinton infomercial. Even before Hillary had a campaign, in 2014, Ready for Hillary chose the Sheraton for its daylong confab with donors.

Hillary started her campaign with a burrito bowl at a Chipotle in Maumee, Ohio, and a promise to "reshuffle the deck" in favor of the middle class. To end it, she chose the Peninsula, where the nineteenth-floor Peninsula Suite, at $25,000 a night, offered a grand piano, velvet armchairs in earthy tones, Italian silk curtains, and a dining room table that seats eight.

One thing the Peninsula didn't offer was easy access to the Javits, which stood on the opposite end of the fat middle of the island, in the muck of construction and gridlock on the far West Side, an area conveniently reached from hundreds of lesser hotels, including the Times Square Sheraton.

When I heard early that morning that Hillary would watch the results come in at the Peninsula, I'd just assumed the five-star hotel was another ballsy move to troll Trump one last time, his anticipated Election Night rage compounded by knowing that out the window a single block down Fifth Avenue from Trump Tower, a victorious Hillary, living as if she too were a billionaire, was leaving her tastefully appointed hotel suite to ride in a motorcade across town as the nation's new president-elect. When I later floated this theory by an aide, she shot it down, explaining simply, "Hillary just loves nice hotels and thought she'd win."

8:34 P.M.

People had compared Elan Kriegel, the mastermind behind the Clinton campaign's data analytics, to John Nash, the paranoid schizophrenic and

Nobel Prize–winning mathematician whom Russell Crowe played in *A Beautiful Mind*. With his husky black eyebrows, five o'clock shadow, and crinkles around his eyes when he smiled, Elan had been the behind-the-scenes beating heart of a campaign that believed it would win, not based on crowd size or voters coming out of their skin each time the candidate spoke but on something more absolute than any of that: math.

His statistical model showed Hillary couldn't lose. Robby said it proved the "enormous amount of room we have to maneuver in the map." Everyone—Democrats, donors, the media—had bowed down in worship of the science and computer-crunching quants who had brought presidential campaigns—once the purview of door-knocking and crusty piles of opposition research and the blind magic of voters' whims—into the modern era.

Until the 8:34 p.m. conference call on Election Night.

"Our models in Florida were off," Elan explained to Hillary's top campaign aides.

The data projections said the surge of early voting and Latinos—the newly arrived Puerto Ricans, for example, drawn into the process by the campaign's brilliant bilingual "Get Out the Vote" effort that reached out to them in churches and nail salons and on the streets blasting reggaeton music from *caravanas*—would give Florida to Hillary. But as the count in Broward County came in, Elan realized the math, that seemingly infallible thing, had been wrong. Turnout among Latinos wasn't what they'd projected, while white voters had stormed the polls in the Panhandle.

Robby, who put his faith entirely in his calculator, had reassured Hillary minutes earlier that she was in good shape. "If we win Florida, we don't actually have to win Pennsylvania," he'd reminded her. On a conference call with senior campaign officials, Robby put on his sunny-bordering-on-delusional disposition. "That's okay. This is all okay. Our path never relied on Florida. We can lose there and still be fine," he said in response to Elan's update.

Long pause. Exhale.

"But, Robby," Elan started slowly, realizing what the next statement would mean, ". . . if our models were wrong in Florida, they could be wrong everywhere."

9:02 P.M.

"Robby, WHAT THE FUCK IS HAPPENING?"

It wasn't an uncommon occurrence for Mandy Grunwald, who'd been in Clinton pressure cookers since 1992, to express her concerns using a choice set of expletives strung together with the élan of a Harvard via Nightingale-Bamford alum. But this time was different.

The campaign had three power centers on Election Night: backstage at the Javits Center, where in between conference calls, aides tried to avoid the press and keep donors and big-name celebrities (including Cher, Lady Gaga, and Katy Perry) calm; the Peninsula, where Robby delivered updates to Bill and Hillary and Chelsea (usually via Huma or Cheryl); and what several aides described as a faux War Room at Hillary's personal office in Midtown, an anonymous block of cubicles and a conference room. That's where Mandy and Joel Benenson, two similarly high-strung New Yorkers whose advice had been largely shunned in the final months of the campaign, and the pollster John Anzalone watched the returns on cable TV and fumed on conference calls at the shocking level of incompetence that slowly, and only late into the night, became apparent. Robby asked Marlon Marshall, his sidekick since the 2008 campaign and the agreeable director of state campaigns who spent most of his time talking to local leaders in Ohio and Pennsylvania and North Carolina, to join Mandy and Joel's frustrated little army to make what was essentially an Election Night rubber room feel a little more like a legitimate campaign operation.

The internal projections had given Hillary a 65 percent chance of narrowly winning Florida. Losing the state and realizing the data had been off panicked everyone. (It was around that time I spotted my friend Ned, who had helped put on the Election Night party and who doesn't smoke, outside the Javits sucking down a Newport menthol he bummed off one of the concession workers.) But for the most part, senior aides remained cautiously optimistic as Elan and Robby tried to figure out whether their modeling had been off only on the East Coast or in the Midwest as well. "There's still a path to victory. We're okay," Robby kept saying.

By the time everyone joined the next conference call, John King's magic wall on CNN began to show Pennsylvania awash in red. Panic spread on the floor of the Javits, where people held cupped palms over open mouths and stood in silence, glued to the giant TV screens set up over the stage.

The only thing worse than the anxiety that now gripped the Javits was the fear that spread across Manhattan Island like a nor'easter and now swept through the modest, virtually empty office on Forty-Fifth Street, where Mandy Grunwald had finally had enough.

9:17 P.M.

Things were already looking bad when Chelsea popped the champagne. In the family's suite at the Peninsula, she was having her hair and makeup done. Charlotte, in a dress decorated with the campaign's *H* logo, napped in an adjacent room. Senior aides ducked in and out, but mostly the serene space was reserved for family and their personal staff, who were like a surrogate family. Chelsea filled everyone's glasses with what somebody told me was Veuve Clicquot, figuring—they all still did—that in a couple of hours Trump's run of early victories in red states (West Virginia, Oklahoma, Alabama) would end, and the map would inevitably turn back in her mom's favor. Hillary had just won New Jersey, and in the minutes before Tennessee was called for Trump, she held a narrow electoral edge. Chelsea wanted to look perfect that night when she would wave to supporters at the Javits; the evening would serve as a pivotal moment in her public evolution from gangly kid to the sophisticated daughter of the first woman president—and the sole heir to the Clinton legacy.

9:47 P.M.

I was out of breath from running up and down the stairs from the Javits Center's sepulchral press area, where hundreds of reporters sat stupefied watching CNN, to the upstairs area where TV crews set up around the perimeter, their correspondents' heads jutting out like gargoyles over the crestfallen supporters below.

I looked around. After Comey's letter and the Anthony Weiner connection, Hillary and Huma agreed it would be best if she worked from Brooklyn for at least a week until the outrage machine moved on. Hillary's friend, Capricia Marshall, pitched in on the road. But the less visible role didn't make Huma any less central to the campaign. She'd poured herself into planning the Election Night festivities, into making Hillary's night impeccable. With advice from *Vogue*'s Anna Wintour, Huma had choreographed the Javits Center festivities with the glitz of a celebrity-wedding planner and the precision of a Broadway stage director. It looked as tasteful as she was, the custom-made stage shaped like the US and blanketed in royal-blue carpet laid for the sole purpose of Hillary, the new president-elect, to walk upon, the two hundred pounds of confetti shaped like shards of glass set to shatter and pour down from the atrium ceiling, the Empire State Building lit up red, white, and blue and glistening in the distance.

Now, the most beautiful party I'd ever been to, a multicultural bouquet befitting the FWP, was starting to feel like a mass funeral. The Nasty Women had become zombified. Their mascara ran; they let out a collective gasp each time the next state turned red. The gay men hugged each other, some collapsing into human comfort blankets on the floor. I'd called in a favor to get my childhood friend Barry Dale into the Javits. He'd been exuberant until about 9:30, when he went home to his apartment, took a Xanax, and didn't wake up until the next morning.

I'd been talking to Andrea Mitchell when a Muslim woman in a hijab dove her torso over the press barricade and grabbed onto my arm. "Tell me she can still win. What are you seeing? Tell me." I couldn't bring myself to tell her that the Upshot needle had swerved away from nearly 90 percent for Hillary earlier in the evening and quickly plummeted toward Trump, giving him a 50 percent, 60 percent, 70 percent chance . . .

Earlier, the Javits had been swarming with campaign aides and Democrats, all happy to do TV hits about how confident they were. But after Florida, almost everyone quotable—even Hired Gun Guy—disappeared.

Brooklyn gathered in a windowless holding room, the kind we'd used for hurricane drills in Texas. Robby and Podesta warned everyone not to talk to the press—as if that mattered now—so the only information I could get came via text.

Sean, the same source who had helped me break the Kaine news, texted me all night with mostly optimistic updates. ("He has no chance in NV," and "Nothing is over until it's over.")

But at 9:47 p.m., I looked down and saw the text. My stomach churned.

"We're fucked unless we hold the Rust Belt," Sean wrote.

I called Carolyn and told her Hillary was going to lose.

11:51 P.M.

> **Subject line: Wisconsin**
> **From: Chozick, Amy**
> **To: Carolyn, Jonathan, Patrick, David, Ian**
> **Date: Tuesday, Nov. 8, 2016 at 11:51 p.m.**
> Not gonna happen for them.
> Gone.

> **From: Ryan, Carolyn**
> **To: Chozick, Amy**
> **Date: Tuesday, Nov. 8, 2016 at 11:55 p.m.**
> Where are you getting that
> From who

> **From: Chozick, Amy**
> **To: Carolyn, Jonathan, Patrick, David, Ian**
> **Date: Tuesday, Nov. 8 at 11:57 p.m.**
> From BK source who is at Javits, says Milwaukee is gone and that was her best shot.

Of all the Brooklyn aides, Jen Palmieri had the most pleasant bedside manner. That made her the designated deliverer of bad news to Hillary.

But not this time. She told Robby there was no way she was going to tell Hillary she couldn't win. That's when Robby, drained and deflated, watching the results with his team in a room down the hall from Hillary's suite, labored into the hallway of the Peninsula to break the news. Hillary didn't seem all that surprised.

"I knew it. I knew this would happen to me . . ." Hillary said, now within a couple of inches of his face. "They were never going to let me be president."

1:36 A.M.

Hillary had already taken to bed when the AP called Pennsylvania for Trump. There had been talk of Chelsea or Bill addressing supporters at the Javits, but that wasn't happening. Podesta agreed to do it. Cheryl Mills was on the phone with the lawyers about possible voting irregularities and recounts. Bill lay on a sofa in an adjacent room, silent and gnawing on the rotting end of an hours-old cigar.

Nobody planned for Hillary to concede that night. They'd decided to fight. But the contract at the Javits was expiring and supporters would be kicked out soon.

"We can wait a little longer, can't we?" Podesta told the crowd. He instructed everyone to go home. "They're still counting votes and every vote should count. Several states are too close to call so we're not going to have anything more to say tonight."

After his brief remarks, Podesta went into the holding room and assured the campaign staff that "we are coming to the office tomorrow and identifying our narrow path forward."

He really thought that. But back at the Peninsula, Hillary was preparing to call Trump.

2:36 A.M.

The AP declared Trump the winner. Huma called Kellyanne Conway. She patched Hillary through. "I have President-Elect Donald Trump for you . . ." Hillary was brief, a simple "Congratulations, Donald," uttered in a half-asleep haze.

Obama had been glacial when Hillary called him minutes earlier. "You need to concede," he told her. "There's no point dragging this out."

I wandered around Midtown with all my luggage. The streets were populated mostly by homeless people, a few remaining XXX-rated video stores, and Hillary supporters, their VIP Election Night passes still hanging around their necks. I headed to the F train, several long blocks over, but halfway there spotted a taxi. I sprinted toward it and thrust my suitcase in the trunk. We were headed down the FDR when I confirmed Hillary had called Trump to concede. I let my editors know.

The taxi driver overheard my calls.

"I'm from Pakistan," he said, his voice breaking. "Is Trump going to send me back? I have a green card. I've been here twelve years."

"I don't know," I said. I looked out the window at the East River. "I don't know anything."

7:55 A.M.
The final alert from the campaign arrived in my inbox:

> **Hillary Clinton to Offer Remarks This Morning in New York City**
> Hillary Clinton will deliver remarks to staff and supporters at 9:30 a.m. on Wednesday, November 9 at the New Yorker Hotel, Grand Ballroom—481 8th Ave NY NY 10001.

I fell asleep on the couch with all my clothes on—jeans and a plaid flannel shirt that flared in a ruffle at the bottom to hide my gut and that I'd worn for three consecutive days. I hadn't turned the TV off after Trump gave his acceptance speech, and the glow of depleted pundits on a CNN news panel filled the room as I nodded off.

At 6:30 a.m. I heard rain outside and the beeping of a trash truck and bolted upright, disoriented and lightheaded with a chalky taste in my mouth from the protein bars and pretzels, the only things I'd eaten the night before. I had a text from a source in the campaign (or what used to be the campaign) saying Hillary would address supporters at the New Yorker Hotel in Midtown later that morning, and I needed to get there "ASAP."

"Pretend you're in the pool. Text me with any problems," he said.

I rubbed my eyes smearing the mascara I forgot I was wearing all over my face. I typed back, "Will do, thx."

I hadn't seen Bobby in over a week. He came out of the bedroom and without turning on the lights unlocked the dead bolt on our front door and plopped the *Times* on the coffee table. The headline stared at me: TRUMP TRIUMPHS: OUTSIDER MOGUL CAPTURES THE PRESIDENCY, STUNNING CLINTON IN BATTLEGROUND STATES.

Bobby had tears in his eyes, his posture slumped over in his white T-shirt and boxer shorts. "I've been waiting for four years, longer, eight, to pick up the paper the day after the election and see your name," he said. "I've," he stopped himself. "We've given up a lot of our lives, all so you could have that front page."

"That's not why I did it," I said, jumping up to brush my teeth and wash my face.

"Then why?" he asked, sitting down on the sagging part of the sofa where I'd slept.

"I don't know, to witness history? To travel the country? To cover one of the most fascinating figures of the twentieth century. Why are we talking about this? I gotta go. She's speaking soon."

"Sure, go. I haven't seen you in weeks, but go. Hillary calls."

I peeked my head from around the hallway and said to the back of his head, "Well, this is the last time . . ."

A PRINTING PLATE, the sliver of silver aluminum that had been spooled around a cylindrical tube like yarn on a loom and used to print the front page of the *Times* the night in 1992 when Bill Clinton won the presidency, sat on the filing cabinet in front of my cubicle. I glimpsed it every time I walked to the printer. WILLIAM JEFFERSON BLYTHE CLINTON: A MAN WHO WANTS TO BE LIKED, AND IS read the headline of the "Man in the News" story by Michael Kelly. Kelly died covering the Iraq War in 2003. The other bylines on the front page, Robin Toner, the first woman to be the national political correspondent at the *Times*, and R. W. Apple Jr. (aka Johnny Apple), the paper's legendary political journalist, had both died of cancer. But they were still alive on that front page, forever intertwined with history.

For three years, every time I felt like I was too much of a chickenshit to last on the beat (*There's a target on your back . . .*), I'd think about the immortality of a byline, of people rushing to newsstands to buy the paper the morning after Election Day. My name among those beneath the six-column banner headline MADAM PRESIDENT.

I hadn't been in the newsroom on Election Night, but the *Times* had sold "Times Insiders" special access to watch coverage unfold. For two hundred and fifty dollars per person, subscribers in a downstairs auditorium got live play-by-play updates from *Times* journalists. Other assorted VIPs got to tour the newsroom and gawk at editors as if they were zoo animals in need of a tranquilizer.

The Upshot's data projections had been so deceiving that, like most newsrooms, we'd hardly had any contingency plan for a Trump victory. ("We got nothing!" an editor was heard yelling across the newsroom.) A color piece that had been envisioned as capturing the white patrons at a dive bar in a Pennsylvania steel town "crying in their beers" after Trump lost was quickly reworked into a front-page story on how white men had delivered for Trump.

By 10:00 p.m. Michael Barbaro and Fleg locked themselves in an office to write the "Trump Wins" news story. The "Hillary Wins" story Pat Healy and I had spent months on; they hammered out its replacement in a couple of hours.

Before I ran out the door, I picked up the paper off the coffee table. I looked at the image, Trump casting his ballot in a dominant blue tie, Jared Kushner by his side, and skimmed over the words ". . . Mr. Trump's unvarnished overtures to disillusioned voters took hold." The bylines read, "By Patrick Healy and Jonathan Martin." But they hadn't written the story. I later learned that the news desk had been so rushed to make print deadlines that they'd put the wrong bylines on Michael and Fleg's story. The eternal life of a byline bestowed on the wrong reporters. I heard later that one editor apologized for the error saying, "Sorry, there's a lot of shit on that front page."

I was still in reporting mode, consumed with covering Hillary's concession speech and my next story. It would take weeks before it sank in that I would never write the FWP story. The *Times'* byline mix-up had been a stroke of gruesome symbolism—a slap alongside the head that everything we thought mattered, didn't.

54

The Morning After

NEW YORK CITY

From: Varun Anand
Sent: Wednesday, November 09, 2016 10:55 a.m.
To: Traveling Press
Subject: HRC
Has departed the Peninsula and is en route the Wyndham New Yorker Hotel at 10:55 a.m.

By the time I got to the New Yorker Hotel, a couple of blocks from the Javits, hundreds of journalists were snaked around Eighth Avenue and across Thirty-Fourth Street. The air was damp and a light drizzle fell. If I'd gone to the end of the line, there was no way I'd make it inside in time to hear Hillary's concession speech. If I cut in, I'd evoke the wrath of an army of sleep-deprived TV correspondents and their crews of haggard cameramen. I saw the same "tight pool" of Travelers—the Wires, one TV crew that shared its footage with the other networks, and a single print reporter—who'd gone to vote with Hillary the day before, assembled in the foyer. I walked inside.

"I'm the *local* print pooler," I told the press wrangler, Sarah.

Ever since our bus arrived in Waterloo, Sarah had been so upbeat and patient with us, but now she appeared dazed, her eyes bloodshot. She gave me a look as though she didn't have the energy to fight and waved me inside.

The eight or so Travelers who made it inside stood in the art deco lobby not speaking. The hotel staff was replacing a flower arrangement in a large slate urn. I hadn't noticed it until I smelled the fresh forsythia, branches brought back from the Flower District where I'd spent so many mornings at *House & Garden*.

Secret Service led us into a ballroom with stacks of folding tables under black tablecloths and banquet chairs. A lazy-eyed Belgian Malinois, led by a burly agent, moped around to sniff our bags. The Travelers sat on the gold-and-scarlet carpet, a loud violent pattern. We unfurled our chargers, thirsting for a power outlet one last time before we returned to our comfortable office existence. We'd already been swept. I asked permission to use the bathroom.

"Will I still be clean?" I asked the agent, one of my last conversations in Trailese.

"I can hand wand you when you get back," he said.

Wanda, our housekeeper, whom we happened to share with Eric and Don Jr., comes on Wednesdays. As I waited in line for the bathroom, I looked down at my texts. "Don't worry Amy," Wanda wrote. "You can come to WH with me." Smiley face with sunglasses, thumbs-up emoticon. Our Polish cleaning lady had become my closest tie to the White House.

Sarah led us inside to a row of a dozen or so chairs, hardly enough to accommodate the pool, much less the hordes of reporters waiting to get inside. Annie texted me that it didn't look as though she or the rest of the pack would get in. I sat thigh to thigh with the Wires, JenEps to my right. The network embeds set up behind us, unfurling their tripods one last time.

For years, I'd walked up the West Side beneath the glare of the red-lit NEW YORKER HOTEL sign that sat on top of the building. During a trip to New York in college, I'd assumed the brown pyramid-shaped edifice had been the headquarters of the *New Yorker* magazine, all forty-three stories of it. Later, I'd rush past thinking the building condemned, a once-grand squat house for homeless people and hookers, certainly not the kind of place that Hillary would rent out for the biggest speech of her political career and the bookend of all my years covering her.

I'd never been inside. The ballroom looked recently renovated, with crystal wall sconces, chandeliers, and white crown molding. But none of these flourishes could mask the room's sad shabbiness. The recessed lighting, the metal vents that exhaled warm, dry air. The dim yellow light and the harried look of the empty stage awaiting Hillary.

Huma and the other aides, who hours earlier popped champagne on the Stronger Together Express, sat in the front row before the stage, which had been thrown up in front of faux Grecian columns and a creased blue curtain and lined with twelve American flags. Capricia and some of Hillary's girlfriends sat cross-legged on the carpet, their shoulders sloped. The junior Brooklyn staff watched from the gold-rimmed balcony above.

I twisted my neck up to the left and made eye contact with some of the press advance kids who devoted themselves to our lunch orders. Their hopes of a White House job dashed, at least for now. Shell-shocked was the word everyone used to describe the mood. I would've said heartbroken, or worse, gutted.

The half dozen Girls on the Bus who made it inside were all in some stage of a breakdown. Glasses on, no makeup, hair pulled back in tangly bird's nests. We comforted each other with pats on the shoulder. Hugs would've been too conspicuous. The emotions hadn't come so much because Hillary had lost but because her defeat had exposed something about our own insecurities as professional women.

"It was the all-female press corps. It was just too much. The country couldn't take it," Lisa Lerer of the AP said.

For me, the breaking point came with the stream of emails and voice mails from editors beckoning me back to the newsroom. I thought I'd done my paper a service by squirming my way inside, but my editors didn't see the point. They informed me that Matt Flegenheimer would be writing about the speech from the office. Why did it not surprise me at that moment that a less experienced (sorry, Fleg) man got the job? I was only six blocks away from the *Times*. I could get there in ten minutes, fifteen with a stop at Starbucks, yet I still felt as torn between the road and the newsroom as I had on the bus in Iowa or wrestling with the Gogo Wi-Fi on the Stronger Together Express. I pleaded with Carolyn.

"Can I please do this speech story? I will make it very quick and then head back to the office."

"No, we need you to dive into how she lost. We can just get the speech from TV," she said.

I'd never defied Carolyn. I lived to make Mamma happy. But what was she going to do at this point? Punish me by putting someone else on the Hillary beat?

"For ten years of my life, I've covered this woman trying to become president," I said, trying not to let my voice crack. "This might be the last speech I ever see her give. I'm not leaving. I'll come back as soon as it's over."

I stood up and watched as Hillary entered the ballroom with her entourage. Bill looked comatose in his purple satin tie. His mouth hung angry and ajar. Hillary had been caked in makeup, but her eyes still looked as though she'd been crying. *You will look happy*, no use this time.

Brooklyn said she always had two versions of an Election Night speech, but that wasn't true. She'd only had one, the one in which she'd dedicated her victory to her mother. Megan Rooney built out the barebones defeat speech. Hillary knew she wanted to say something about how "we have still not shattered that highest and hardest glass ceiling."

People had all kinds of theories about why Hillary wore purple. The purple of the suffragette flag. The color of LGBTQ anti-bullying. Hillary later said she picked the suit to symbolize red states and blue states that together make purple. She planned to wear it to Washington that day as president-elect. But I had another theory. In Methodism, purple is worn during Advent and Lent, a symbol of penitence. We were watching Saint Hillary.

"I know how disappointed you feel, because I feel it too and so do tens of millions of Americans who invested their hopes and dreams in this effort," Hillary said. "This is painful and it will be for a long time."

I got an email from another (male) editor saying, "Don't use this quote, it's going in Pat's story . . ."

> To all the little girls who are watching this, never doubt that you are valuable and powerful and deserving of every chance and opportunity in the world to pursue and achieve your dreams.

"Funny," I wrote back.

Hillary ended the speech with scripture, Galatians 6:9: "Let us not grow weary, let us not lose heart, for there are more seasons to come and there is more work to do."

AFTERWARD, THE TRAVELERS did what we'd done for so many months—we rushed to the rope line where Hillary greeted the crowd. But this time there was no eruption of selfies. No shouts of "Madam President" and Hillary's thumbs-up replies, "Doesn't that sound good?" and "Let's make it happen!" No "Fight Song" blasting from the speakers. It was silent and grim, like rushing the casket at a wake. Almost everyone was red faced and teary. Hillary extended her arms and leaned in for hugs in a sleepy, habitual motion.

I saw Brown Loafers in his usual position, behind the rope line, close to Hillary. But instead of doing battle with me or trying to protect Hillary ("Watch out for Dan Merica, center right . . ."), he was crying. Not just crying, but sobbing, bellowing until his face was drenched, his whole body convulsed. That was when it hit me. What had it all been for?

For three years, I'd been fighting with The Guys. I'd let my hero of a husband down. I'd put off having a baby. I'd thrown punches in the Steel Cage Match and gained at least twelve pounds. I'd even become an unwitting agent of Russian intelligence.

In the end, we all lost. I was done.

I looked at Brown Loafers that morning and any anger I had dissipated. I had only empathy and respect. I regret that it took Hillary losing for me to see The Guys this way.

I sleepwalked back to the newsroom. The drizzle had turned into a steady rain. I'd been so out of it that I didn't see that the walk light had changed. The screech of an oncoming taxi pulled me back. I did go ahead and stop at Starbucks. As the new president-elect would say, what the hell did I have to lose?

In her 1993 "Politics of Meaning" speech in Austin, Saint Hillary, whose father lay dying in a Little Rock hospital, quoted the late Republican strategist Lee Atwater after he'd been diagnosed with brain cancer. "You can acquire all you want and still feel empty," Hillary said. "What power wouldn't I trade for a little more time with my family? What price wouldn't I pay for an evening with friends?"

I had that quote in my head as I lingered in the Times building's glass-enclosed lobby. I didn't know what my life would be like or who I'd be without Hillary. Hell, I didn't know what my days would be like without being ushered between several swing-state cities (*You are in Tampa . . .*), a flight attendant handing me slunch at 3:15 p.m., and checking into a Hampton Inn. I didn't know what would happen to all the Nasty Women, especially the older women I'd met all over the country, who now doubted they'd ever see a FWP.

I hadn't been in the newsroom in weeks. I dug around in my backpack, looking through a multicolored tangle of credentials for my *Times* security badge.

"Sorry," I told the guard. "I must have left my badge upstairs. I've been on the road, on the campaign trail, for a while."

"No problem," he said. "Welcome back." And, thirteen years after security led me out of the Times building after that political reporter had stood me up for coffee to talk about my career, the guard smiled and pushed the glass gate of the red-lacquered turnstiles open.

"Thank you," I said and walked inside.

The commotion in the newsroom all around me seemed to be happening in slow motion. Everyone was huddled around the life-size Taylor

Swift cutout, looking as shell-shocked as the Brooklyn staff had been, contemplating what we'd done wrong, how we'd missed it.

"God, I didn't go to a single Hillary or Trump rally, and yet, I wrote with such authority," a colleague said.

I sat down at my cubicle to write the "How She Lost" story. Then I finally cried.

I never imagined I'd be a political reporter at the *Times*. That I'd be paid to travel the country—forty-eight states when it was all over—to cover a presidential candidate. That I'd write front-page stories. That I'd be important enough to be called a cunt on Twitter daily or to sometimes get the coveted aisle seat. And the truth was I couldn't have done it without Hillary.

Not because writing about her raised my profile. (Though, I do realize the Carly Fiorina beat wouldn't have led to a lot of A1 stories and a book deal.) But because no one else could fascinate and inspire and infuriate me all at the same time the way Hillary could.

SIX MONTHS AFTER the election, I still dream about Hillary. At my checkup last week, Dr. Broderick warned me that my dreams will get more vivid now that I'm pregnant. I didn't tell her that the night before, Hillary and I had been riding in a press van to a Trump rally, crammed into the back seat on either side of Justin Trudeau.

I keep going back to what Hillary said when she found out she was pregnant with Chelsea. They'd tried for several years and were on their way to California to see a specialist when she heard. Hillary, then the first lady of Arkansas, was adjusting to infidelity and the boom-and-bust cycles of life with Bill Clinton. She went to a girlfriend's house to share the good news. The two women sat on a patio in Little Rock's leafy Hillcrest district. They sipped iced tea. "Oh, I'm just so happy," Hillary said. "For the first time in my life, I don't have to *do* anything. My body will do everything for me."

That is the Hillary I want our child to understand, not the historical figure who lost to Donald Trump in a very strange and ugly election in the year 2016. But the Hillary who spent her life *doing*. The Hillary who tried to hold it all together—her marriage, her daughter, her career, her gender, her country. The Hillary who taught me about grit. Who showed me how to revolt against the dunces, all in confederacy against me. To believe I could infiltrate the elite media. To remember that you can acquire all you want and still feel empty, devote yourself so entirely to something and still fall short. To accept that I could finally stop striving.

Hillary taught me all of that. So what if she hated me?

The author and Hillary Clinton, Iowa, 2007

Acknowledgments

"Put it in the book!" That's what Carolyn Ryan would say anytime I told her something juicy that wasn't quite right for the paper. There would be no book without her—or the *Times*. She even came up with the title.

Dean Baquet always had my back on the Hillary beat and was also an early, enthusiastic supporter of this book. Matt Purdy and Alison Mitchell's love of journalism always made me psyched to be a part of even (or especially?) the most stressful stories. The late Janet Elder knew when I needed a pep talk or a shoulder to lean (or cry) on. We miss her. When Jill Abramson hired me, she told me to think of the *Times* as the "ultimate buffet" for a writer—and that is exactly what it has been to me. I will always be grateful to Jill for putting me on the Hillary beat and for her friendship.

Only a handful of people knew about this book when I first wrote the proposal in 2014. I am blessed that they turned out to be the right people.

David McCormick was as much an editor as an agent, helping me shape the idea and offering me invaluable advice and encouragement throughout every step. Jonathan Jao transcended the role of book editor. For years, every time I doubted that anyone would want to read my story, I'd remember his giddy enthusiasm and vision for what this book could be. I leaned on his intellect, wry wit, and endless patience for even the smallest of word choices. Jonathan Burnham and Doug Jones at HarperCollins were an author's dream, keenly interested in my idea from the start and remaining devoted throughout the 2016 election and beyond. I'd hardly spit out my vision for a memoir when Tina Andreadis was scheming up ways to promote the hell out of it. In addition to being a publicist, Tina has been a trusted friend, therapist, and savvy reader (#InTinaWeTrust). Amanda Pelletier, Sofia Groopman, and Emily Taylor at HarperCollins should really be running the world one day—or at least the publishing industry. I'm thankful to have Kassie Evashevski in my corner. Hiring Benjamin Phelan, a tireless fact-checking machine, was the best money I ever spent. (Thanks to Katy Tur for connecting us.)

I am immensely grateful to Hugo Lindgren, a magazine-editing virtuoso who published my "Planet Hillary" story and, three years later, helped me with this book. Dwight Garner and Louise Story both gave an early draft a close, careful read and provided essential feedback. Martin Wilson, a true friend and novelist, inspired me to keep my own ass in the chair. A big thank-you to Julie Bosman, Tracy Sefl, Rich Turner, Julie Bloom, Stephanie Clifford, Jeffrey Gettleman, Jon Kelly, and Risa Heller, who offered me their time and wisdom on everything from the prose and the PR to the jacket design.

Several *Times* editors, including David Halbfinger, Ian Trontz, and Gerry Mullany, masterfully edited many of the stories referenced in the book. Bruce Headlam first hired me to be a media reporter at the *Times* and has been a friend (and editor) to me ever since. Bill Brink is like my office dad. Ellen Pollock is a longtime role model. Mark Leibovich helped me navigate balancing book and newspaper writing, even as he made it look easy. I have too many brilliant, witty, warm *Times* colleagues to thank everyone, but Jeremy Peters, Michael Barbaro, Maggie Haberman, Patrick Healy, Maureen Dowd, Michael Schmidt, Peter Baker, Jonathan Martin, Jason Horowitz, Dagny Salas, Michael Gold, Jessica Dimson, Juanita Powell-Brunson, Nicholas Corasaniti, Brooks Barnes, Sue Craig, Kitty Bennett, Steve Eder, and Trip Gabriel come to mind. Tom Kaplan and Matt Flegenheimer made every day of the campaign (even those grueling "Death March" months) more fun.

Thank you to the rest of Hillary's traveling press on both of her presidential campaigns. We were assigned to the same candidate's bus by chance, but many of you—too many to name—became dear friends and my de facto traveling family.

I also want to express heartfelt appreciation to my sources (mostly, but not all, women) who had the moxie to defy Hillary's press handlers and talk to me anyway. I can't name them (for obvious reasons) but I can say that without their trust, insights into Hillary, and patience with my endless phone calls and follow-ups, I wouldn't have been able to do my job or write this book.

The best for last . . .

This book is devoted to my family, who has always given me unconditional support and was ready to place their preorders before I wrote a word.

Sandra and Fred Kline, Lis and Fred Chozick, and Gary and Jessie Jacobs haven't missed a single important event in my life. From the start, Lisa and Jeff Blau have been excited about this book and eager to help in any way they could. My Irish family: Fionnuala, John, David, Bryan, and Aisling Ennis—having you in my life has been a gift. Barry Dale Johnson, whose friendship hasn't wavered since we first bonded over Mötley Crüe in the sixth grade, might as well be family. Grandma Rose may have taken me to Vegas, but I wouldn't be who I am without the love and influence of my mother's late parents, Ada and Milford Jacobs (aka Gummy and Guppy). My sister, Stefani Shanberg, is a hilarious, bad-ass, best friend whom I look up to and credit with toughening me up. (Thanks to my most terrific brother-in-law, Dave Shanberg, too.) I joke with my parents, Ronni and Jason Chozick, that maybe if I had a more tortured childhood, they would've been more prominent characters. But as it turns out, they were the absolute best, most loving, supportive parents any kid could possibly hope to have. They've been eager to read and gush over this book (indeed, any book I wrote) ever since my debut at the Young Author's Conference in the first grade. But they would've been proud of Stef and me no matter what direction our lives took. That's the kind of parents they are. I can't begin to thank them enough.

If you've gotten this far, you already know what kind of husband Bobby is, so I won't get into the details here except to say that not a word of this book would exist without him. He is everything.

Finally, to our son, Cormac, whose tiny kicks sustained me as I wrote this book and who graciously entered this world just before deadline. You are the love of our lives.

About the Author

AMY CHOZICK is a writer-at-large for the *New York Times*. Originally from San Antonio, Texas, she lives in New York with her husband and son.